AF426141

PLC HARDWARE & PROGRAMMING

Revision 1.0
ISBN 979-8-218-28414-5

Allen-Bradley Platforms

By

Frank Lamb

© 2023 Frank Lamb, Automation Consulting, LLC. All rights reserved. Except as permitted under the United States Copyright Act of 1976, no part of this publication may be reproduced or distributed in any form or by any means, or stored in a data base or retrieval system, without the prior written permission of the publisher.

Information contained in this work has been obtained from sources believed to be reliable, however, the author shall not be responsible for any errors, omissions or damages arising out of use of this information. This work is published with the understanding that Automation Consulting, LLC and the author are supplying information but are not rendering engineering or other professional services. If such services are required, the assistance of an appropriate professional should be sought.

Parts of this book were extracted from the book "Advanced PLC Hardware & Programming" © Automation Consulting, LLC, October 2019, Rev.1.1

www.automationllc.com

This Page Intentionally Left Blank

Table of Contents

Introduction

This book was originally written as part of "Advanced PLC Hardware and Programming" in 2019, which in turn evolved from the original "PLC Hardware and Programming – Multi-Platform", published in 2016. In 2022, after taking on a contract to develop training materials for Automation NTH, PLC Hardware and Programming was re-published under my Automation Consulting, LLC label as part of the NTH University training program.

NTH University is an engineering training program for interns, apprentices, and employees of Automation NTH (www.automationnth.com). Automation NTH is a control systems integrator located near Nashville, Tennessee. They also provide turnkey manufacturing and machinery solutions through their Certified Automation Partner (CAP) program. The NTHU classes are sometimes also offered to outside customers, so making separate platform books for each brand that augment the Multi-Platform book makes sense.

The platform-specific books will contain many of the advanced concepts and exercises found in the Advanced PLC Hardware and programming book, written specifically for each software.

- Frank Lamb, April 2023

PLC Hardware & Programming - Overview

The original book approached PLC training from a generic viewpoint. Most PLC platforms have many things in common; before beginning the study of a particular brand of PLC, it is important to learn the things that are common to *all* platforms. The book covers general topics like hardware and layout, bit logic, timers, counters, math and troubleshooting. It is written primarily in Ladder Logic, which may not always be appropriate.

The book begins with a history of computing devices, including the evolution of the PLC. It then describes a PLC's physical layout, discusses I/O wiring and devices, covers various forms of communications with an emphasis on ethernet, then discusses data types and numbering systems. The five IEC 61131-3 programming languages are described, followed by program organization and scanning.

The rest of the book contains a tutorial on ladder logic in a generic format. Bit logic, Timers, Counters, Math and Comparisons and file data treatment are discussed in a generic way that applies to all PLCs. There are also some examples of communication setup for Allen-Bradley and Siemens PLCs.

There are 12 exercises in the book. All are intended to be answered in a generic way, generally in a written format.

How to Use this Book:

The intent of these platform-specific books is to serve as an advanced add-on to PLC Hardware and Programming. It is not necessary to purchase the original book if you are already familiar with PLCs, but the author uses both books together to teach both beginning and advanced platform-specific classes in person.

Communications for Allen-Bradley PLCs uses a separate program called **RSLinx**. This is discussed before the programming sections.

This book covers both the older **RSLogix 500** software which programs the MicroLogix and SLC PLCs, and the newer **RSLogix Studio 5000** which programs the CompactLogix and ControlLogix. The RSLogix 500 software can only be used to program in Ladder, though logic can be viewed and edited in Instruction List, which Allen-Bradley calls ASCII Mnemonics.

RSLogix Studio 5000 allows programming in Structured Text (ST), Function Block Diagram (FBD), and Sequential Function Chart (SFC), however these are not covered extensively in this book. As with the 500 platform, ASCII Mnemonics can be used to modify logic.

As a review of the basics of PLC programming, the original exercises in the PLC Hardware and Programming book are completed in the RSLogix 500 section. This is followed by a more advanced treatment of programming and organization techniques, programming tips and tricks, and a section on simulation.

A Note on Templates and Standards

A **PLC template** is a partially completed program with code examples and comments instructing the programmer. It has guidelines on program structure and sample code inside of the different routine types, and serves as a starting point for a program.

A **PLC programming standard** is a document establishing programming styles and rules. It works together with a template to guide the programmer so that the finished program is readable and easy for technicians or other programmers to follow.

There are well-established software standards and procedures for application programming for languages like C#, Python or Java, but no universal standard for PLC programming. One reason is that every manufacturer's platform is different, even within the same company as you will find with Logix 500 and 5000. This book should be very helpful if you are writing a template or standard for Allen-Bradley platforms, examples are typical for large machine builders, system integrators and manufacturers.

Communications - Allen-Bradley RSLinx

All online activities for Allen-Bradley products use a communications program called RSLinx. Before using the program to perform activities, such as uploading, downloading or going online, RSLinx communications drivers need to be configured. To open RSLinx, you can either browse for it in the Programs list on your computer or simply open the PLC program using RSLogix.

 When a program is opened, RSLinx will often open automatically as a service. The icon to the left is the RSLinx service. You can double-click on the icon to open it or browse for it in the Programs list.

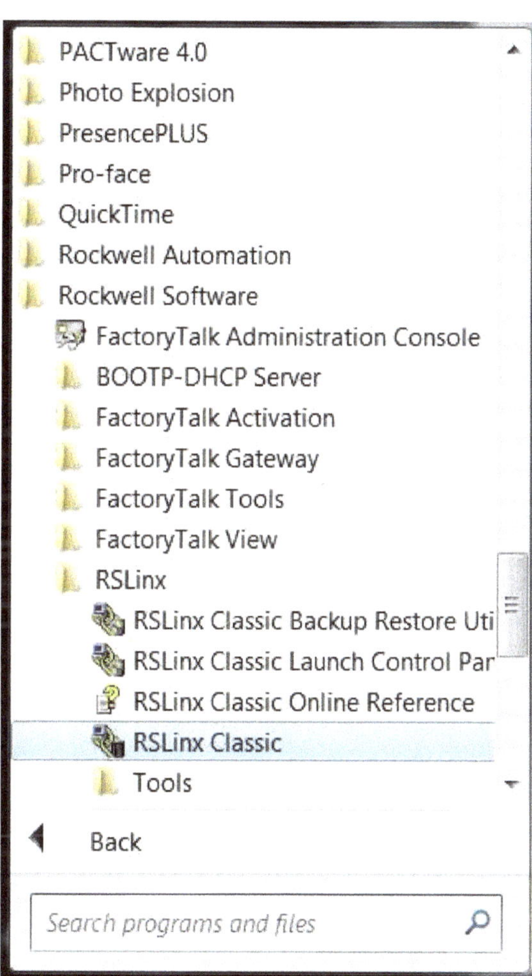

Drivers may already be configured in RSLinx. If they are not, the following procedures can be used.

Newer Allen-Bradley PLCs have several methods of communications, including USB, RS232 DF1 (Serial Communications) and Ethernet. USB communications is used with the ControlLogix L70 series of processors; the port is on the front of the processor.

Serial communications is available on all other processors. They either use a cable with a 9-pin plug on one end and a round plug on the other (MicroLogix), or a 9-pin serial cable with a null modem adapter, as shown in the RS-232 diagram in this document's Communications section. Ethernet uses a standard non-crossover Ethernet cable if using a switch (normal) or a crossover cable if attaching directly to the processor or Ethernet card. The communication ports are on the left side of the PLC for MicroLogix, on the processor for the SLC 5/05 or CompactLogix, or on the bottom of the Ethernet card if using a ControlLogix system.

When a PLC is first commissioned, it has no Ethernet address assigned to it. The Ethernet address can either be assigned using the BOOTP Server utility from Allen-Bradley or by downloading the program using the serial cable. If the Ethernet address has been set up in the PLC program, a serial download will also configure the Ethernet port.

Serial Driver:

1. Open RSLinx and select Communications > Configure Drivers.

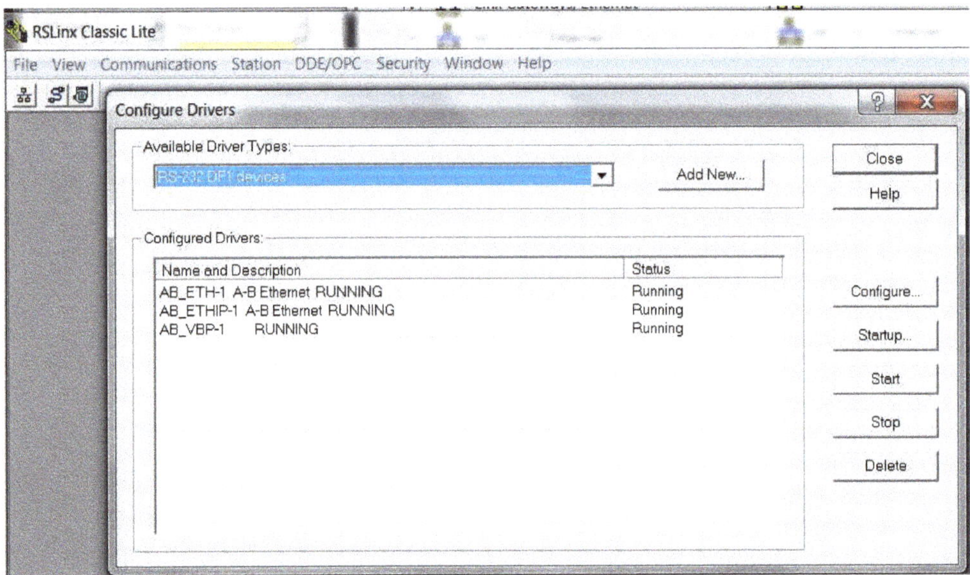

2. Select **RS-232 DF1** devices in the pull-down menu under Available Driver Types. Press the "Add New" button. A dialog box will appear asking you to name the new driver. The default is AB_DF1-1; press the OK button to keep the default.

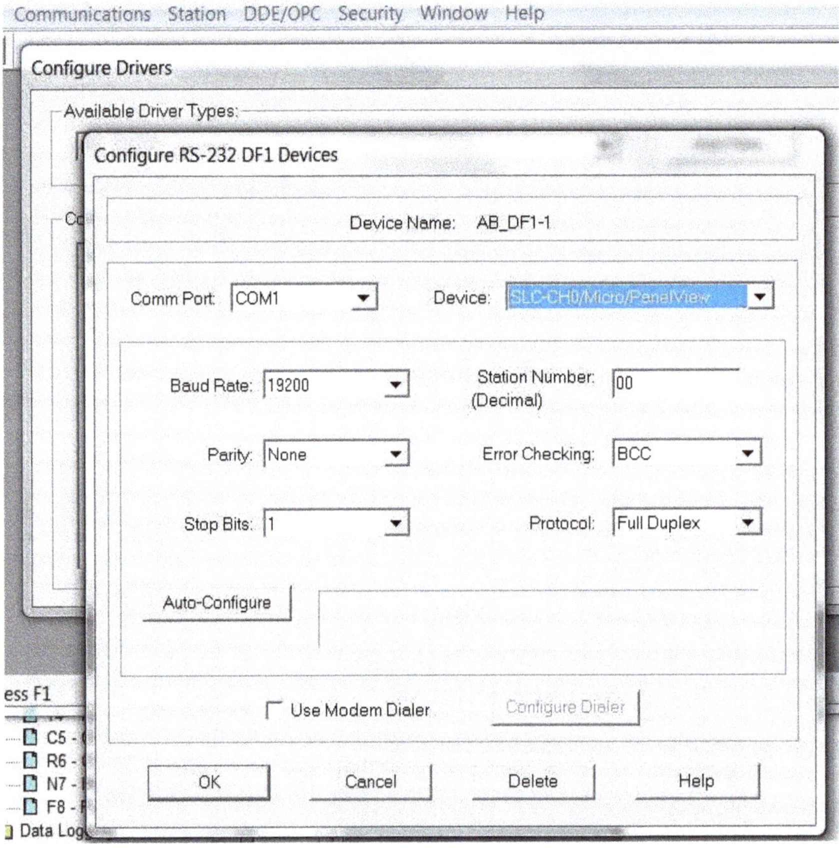

3. Under the Device pull-down, select SLC-CH0/Micro/Panelview (SLC or MicroLogix) or Logix 5550/CompactLogix. Other settings are as shown in the above diagram except for Error Checking, which should be set to CRC.

If the cable is connected to the processor and it is powered up, you can press the Auto-Configure button. This should interrogate the PLC for its current settings and you will receive the message "Auto Configuration Successful!" This confirms that you have communications with the PLC and are ready for download.

Because most computers no longer have a serial port, it is often necessary to use a USB to Serial adapter. Some of these adapters will assign a serial port automatically; if so, you will need to use Device Manager on your computer to discover which port is used. If you see no assigned port for the adapter, try Port 8 or higher and press Auto Configure. Even if the driver displays "Port Conflict", it will usually show the "Auto Configuration Successful!" message anyway.

It is often necessary to use a "Null-Modem" adapter on the serial cable. This crosses pins 2 and 3 (TX and RX). This is true for ControlLogix and SLC500 processors.

To Configure the Ethernet Drivers:

There are two Ethernet drivers listed in the Available Driver Types list. The first is "Ethernet Devices" and the second is "Ethernet/IP Driver". **Ethernet Devices** allows the computer to locate processors and other devices by typing in the address; it works on all Allen-Bradley Ethernet items and also finds many compatible devices not made by Allen-Bradley.

Ethernet Devices:

1. Open RSLinx and select Communications > Configure Drivers.

2. Select Ethernet Devices in the pull-down menu under Available Driver Types. Press the "Add New" button. A dialog box will appear asking you to name the new driver. The default is AB_ETH-1; press the OK button to keep the default.

If there is more than one group of Allen-Bradley processors and devices in your plant, it may be advisable to create more than one Ethernet Devices driver and give them different names, such as ETH_Line1, ETH_Line2 etc. This is to prevent the driver from attempting to find devices that are not present, which takes more time.

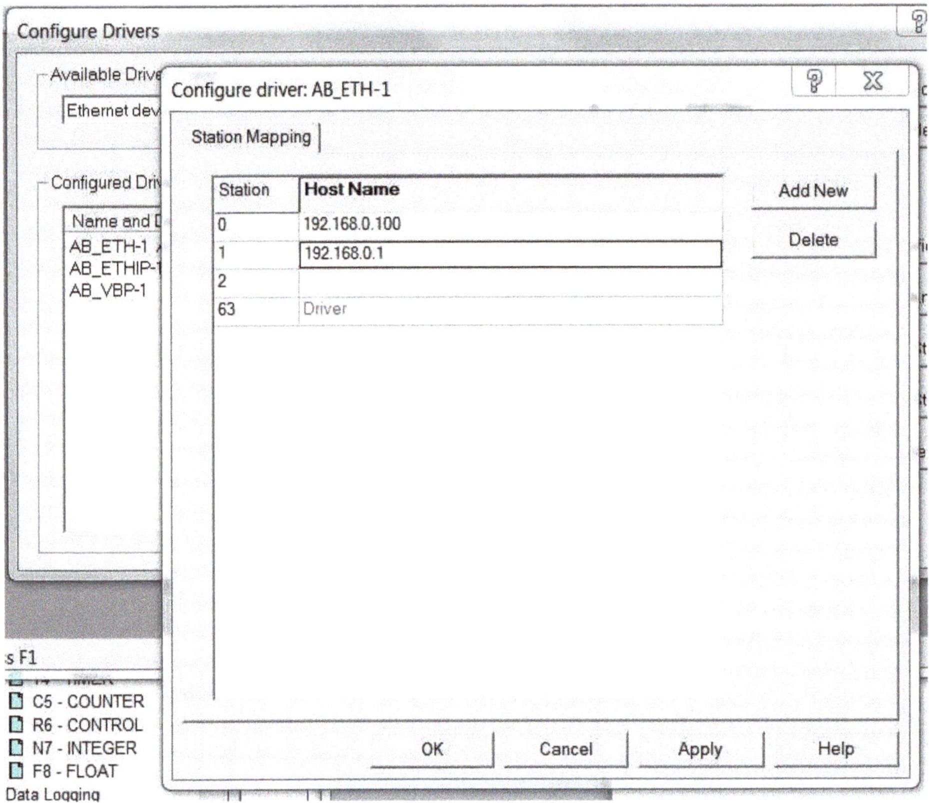

3. Type in the Ethernet addresses as shown. This assumes a computer with assigned address 192.168.0.100 and a PLC at 192.168.0.1. The first address (Station 0) is not necessary to communicate with the PLC, but it is there to show the current configuration of the computer itself. The second address (Station 1) is the address configured for the communications port in the PLC program. More PLCs and devices can be added to this list as required.

Ethernet/IP Driver:

This driver works for all Allen-Bradley CIP (Ethernet/IP) devices, which does not include the SLC 5/05 and MicroLogix. It does not require that you know the addresses of the devices.

1. Open RSLinx and select Communications > Configure Drivers.

2. Select EtherNet/IP Driver in the pull-down menu under Available Driver Types. Press the "Add New" button. A dialog box will appear asking you to name the new driver. The default is AB_ETHIP-1; press the OK button to keep the default.

3. This driver only requires that you select your Ethernet card from the list of devices on your computer. This diagram shows a wired port (192.168.4.204) and a wireless card (50.94.219.161)

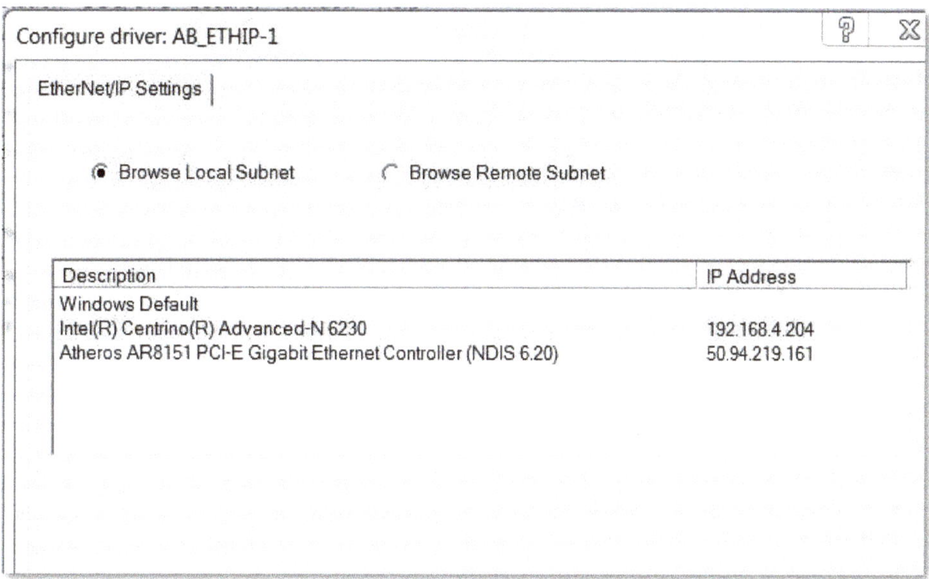

Since this driver does not maintain a list of devices, it is not necessary to create multiple drivers as with the Ethernet Devices driver. Previous connections may show up on the list when browsing in "RSWho"; just right click on them and select "remove" to eliminate them.

Allen-Bradley PLCs

Allen-Bradley traces its origins to 1903 and the formation of the Compression Rheostat Company, founded by Lynde Bradley and Dr. Stanton Allen with an initial investment of $1000. In 1904, 19-year-old Harry Bradley joined his brother in the business, and in 1909, the company was renamed the Allen-Bradley Company.

The company expanded rapidly during World War I in response to government contract work. Its product line at that time included automatic starters and switches, circuit breakers, relays and other electrical equipment. Located in Milwaukee, the first sales office was established in New York.

During the 1970s, the company expanded its production facilities and became a global company. In 1985, the company was purchased by Rockwell International.

In 1994, Rockwell Software was launched. Rockwell also acquired Reliance Electric and Dodge; the collection of companies was marketed as Rockwell Automation.

In 2002, Rockwell International split into two companies. The industrial automation division remained as Rockwell Automation, while the avionics division became Rockwell Collins.

The original large rack PLC system was the PLC, followed by the PLC2, PLC3 and PLC5. The PLC5 is still supported.

In 1991, the SLC (Small Logic Controller) made its debut as a smaller version of the PLC5, using an abbreviated version of the PLC instruction set. The MicroLogix family appeared in 1995, followed by the first ControlLogix tag-based platform in 1997.

Rockwell Software still makes all the programming and communications software for the Allen-Bradley PLC and PAC (Programmable Automation Controller) families.

Allen-Bradley SLC and MicroLogix Platforms

Rockwell Software – RSLogix 500

The main software package for programming the SLC and MicroLogix families of PLCs is
RSLogix 500. The current release as of June 2017 is version 11.0.

Name	Catalog #	Controllers	Description
Professional Edition	9324-RL0700NXENE	All SLC 500 and MicroLogix controllers except 800 series	Online/Offline Programming, Includes RSNetworx for ControlNet/ DeviceNet/ Ethernet/IP and RSLogix Emulate
Standard Edition	9324-RL0300ENE	All SLC 500 and MicroLogix controllers except 800 series	Online/Offline Programming
Starter Edition	9324-RL0100ENE	All SLC 500 and MicroLogix controllers except 800 series	Offline Programming only, no cross referencing, Data usage, Program Compare.
Micro Developer	9324-RLM0800ENE	All MicroLogix controllers except 800 series	Online/Offline Programming
Micro Starter	9324-RL0300ENE	All MicroLogix controllers except 800 series	Online/Offline Programming, no Data usage, Program Compare, Trending, Advanced Diagnostics, Program Report.
Micro Starter Lite	Free	MicroLogix 1000 and 1100 only	Online editing for MicroLogix 1100 only, no Data usage, Program Compare, Trending, Advanced Diagnostics, Program Report.

The Micro800 family of controllers is programmed using the Connected Components
Workbench, which is also used for other devices such as VFDs, Servo Drives, Light Curtains and
Safety Relays.

Name	Catalog #	Controllers	Description
Developer Edition	9328-CCWDEVENE	All Micro800 products	Offline programming, Run Mode Change, Spy List, UDT, IP Protection
Standard Edition	Free	All Micro800 products	Offline Programming

MicroLogix 800

Software:

Connected Components Workbench, (Standard Edition is free). Ladder, FBD, Structured Text support.

Models:

Micro810: 12 I/O points with 4 high-current relay outputs, DC models allow 4 inputs to function as 0-10v analog inputs, Real Time Clock, Optional 1.5" local LCD for monitoring/modifying application data.

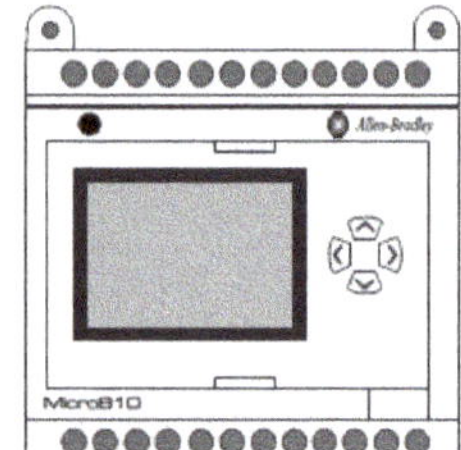

Micro820: Up to 36 I/O points, up to 2 plug-in modules, Ethernet, MicroSD card for recipe or data logging, 5khz PWM output, Real Time Clock, Optional 3.5" LCD Display, analog I/O.

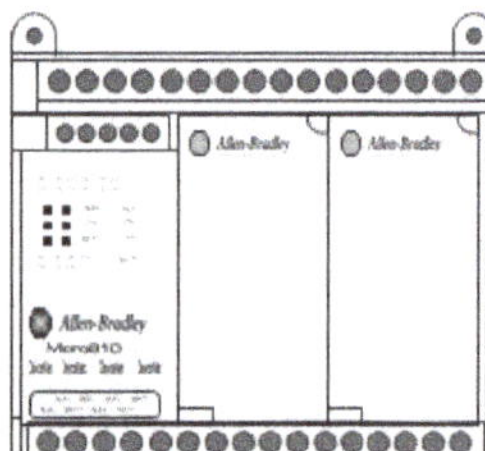

Micro830: Up to 88 I/O points with up to 20 analog inputs, Hi performance I/O, Interrupts, PTO Motion, 2080 expansion I/O.

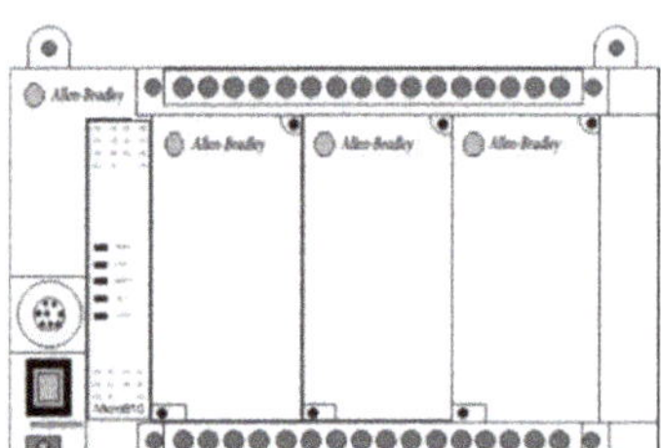

Micro850: Up to 132 I/O points, Hi-performance I/O, Interrupts, PTO Motion, embedded Ethernet, 2085 expansion I/O. Can function as RTU unit for SCADA (Serial Modbus or Ethernet).

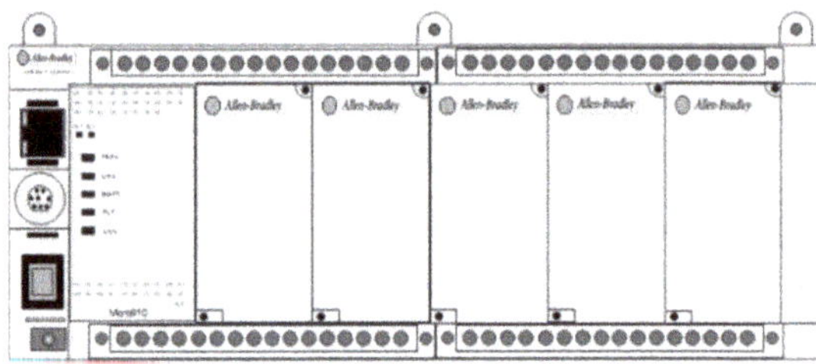

MicroLogix 800

MicroLogix 1000

Software:

RSLogix 500, RSLogix 500 Micro. Ladder Logic only.

Models:

1763-LXX, where XX denotes the number of I/O points. AC and DC power supplies, I/O ranges from 10 to 32 points. AC or DC Inputs; AC, DC and Relay Outputs. Analog available on some models.

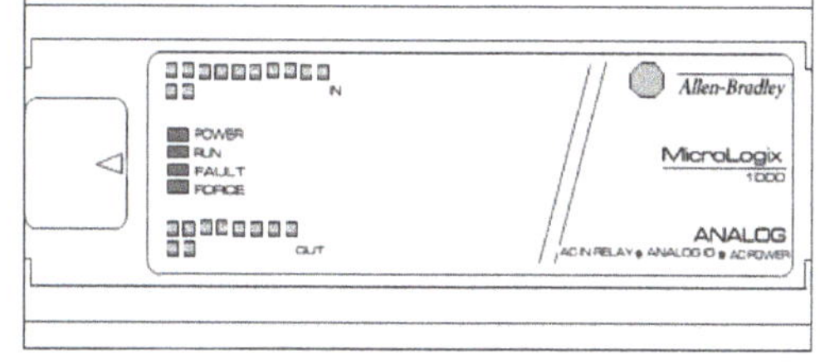

The MicroLogix 1000 is discontinued as of June 30, 2017. Recommended replacement is the MicroLogix 820.

MicroLogix 1100

Software:

RSLogix 500, RSLogix 500 Micro Starter Lite. Online Editing. Ladder Logic only.

Models:

1763-LXX, where XX denotes the number of I/O points. AC and DC power supplies, I/O ranges from 10 to 16 points. Expandable to 144 I/O. AC or DC Inputs; AC, DC and Relay Outputs. Analog available on some models. Ethernet/IP, DH-485 and Modbus RTU capable.

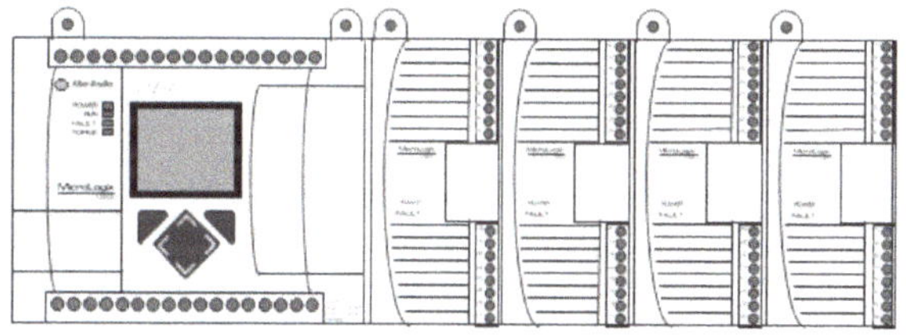

MicroLogix 1200

Software:

RSLogix 500. Ladder Logic only.

Models:

1762-LXX, where XX denotes the number of I/O points. AC and DC power supplies, I/O ranges from 24 to 40 points. Expandable to 136 I/O. AC or DC Inputs; AC, DC and Relay Outputs. Analog available on some models. RS232, RS485 Combo Port. 2 built-in potentiometers.

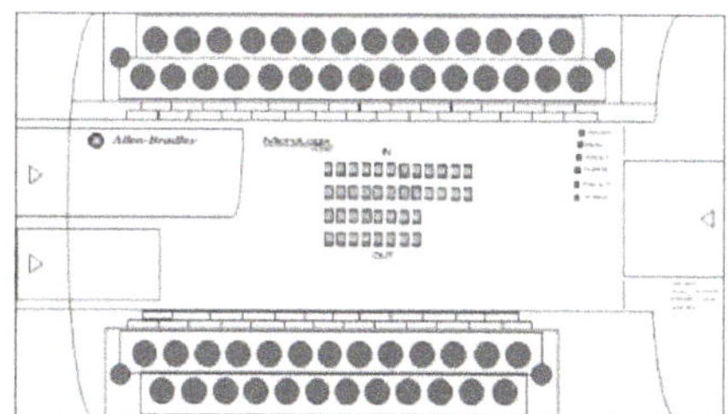

MicroLogix 1400

Software:

RSLogix 500, **RSLogix Micro.** Online Editing. Ladder Logic only.

Models:

1766-LXX, where XX denotes the number of I/O points. 10K words user program memory, 10K words user data memory

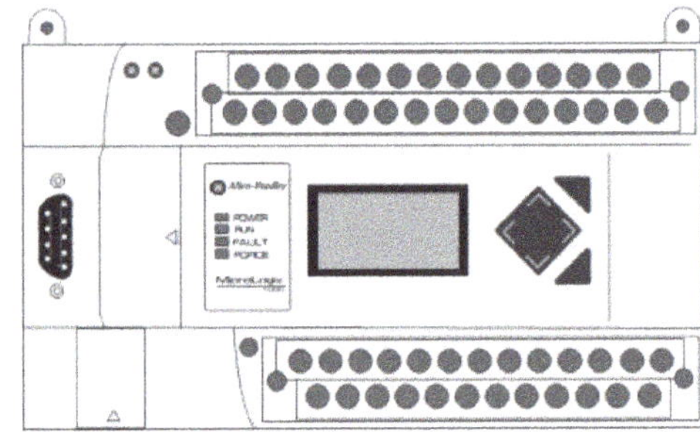

Built-in LCD, AC and DC power supplies, 32 built-in I/O points. Expandable to 256 I/O (7 modules). AC or DC Inputs; AC, DC and Relay Outputs. Analog available on some models. Two built-in serial ports, (DF1/DH485/Modbus RTU/DNP3/ASCII); Ethernet port (EtherNet/IP, Modbus, DNP3). Memory cards for data logging (128K) or recipes (64K).

MicroLogix 1500

Software:

RSLogix 500. Ladder Logic only.

Models:

1769-LXX, where XX denotes the number of I/O points. Built-in LCD, AC and DC power supplies, 32 built-in I/O points. Expandable to 512 I/O. AC or DC Inputs; AC, DC and Relay Outputs. Analog available on some models. Two built-in serial ports, Ethernet port.

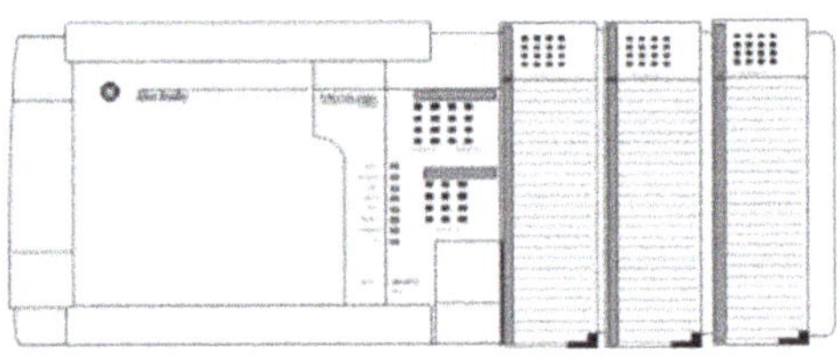

The MicroLogix 1500 is discontinued as of June 30, 2017. Recommended replacement is the MicroLogix 1400 system or the CompactLogix 5370 L1 and L2.

SLC500 Series

Software:

RSLogix 500. Ladder Logic and Structured Text. Advanced instruction sets includes file handling, sequencer, diagnostic, shift register, immediate I/O and program control instructions.

Models:

1747-LXXX, where XXX denotes the family of controllers. The SLC (Small Logic Controller) family is a rack-based system, though the older SLC150 and SLC500 controllers were fixed and had on-board I/O. These have been discontinued.

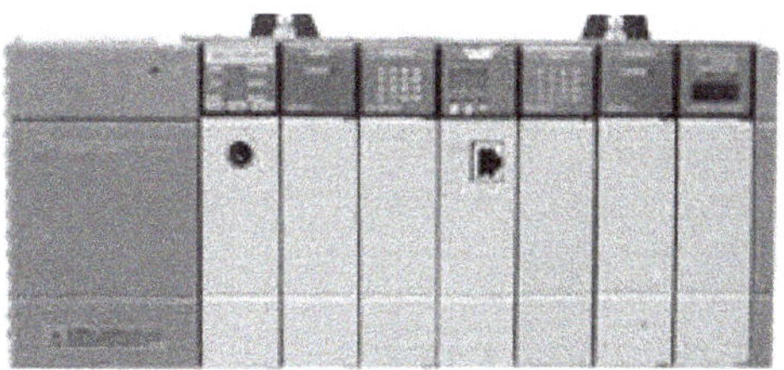

Processors reside in the leftmost slot (slot 0) of the rack, which comes in 4, 7, 10 and 13 slot sizes. Processors include the 5/01, 5/02, 5/03, 5/04 and 5/05. The 5/01 and 5/02 processors have been discontinued, and the 5/03 8K series will be discontinued as of August 31, 2018. Recommended replacements are the CompactLogix 5370 and 5380 series.

Processor	Communications	Memory	Max I/O	# Instructions
SLC 5/01	DH-485 Slave	1K, 4K	3940	51
SLC 5/02	DH-485	4K	4094	~70
SLC 5/03	DH-485, DF1	8K, 16K	4094	~100
SLC 5/04	DH+, DF1	16K, 32K, 64K	4094	~100
SLC 5/05	Ethernet, DF1	16K, 32K, 64K	4094	~100

Online editing is possible with the 5/03, 5/04 and 5/05 processors.

SLC and MicroLogix Memory Registers

Structure: <File> : <Element Number> <Delimiter> <Bit or Word Number>

File	Number	Name	Example	Description	Comment
O	0	Outputs	O:5/3	Fifth Slot, Fourth Bit	Physical Digital Outputs
			O:6.1	Sixth Slot, Second Channel	Physical Analog Outputs
I	1	Inputs	I:4/6	Fourth Slot, Seventh Bit	
			I:7.3	Seventh Slot, Fourth Channel	
S	2	Status	S:1/15	First Scan Bit	Can't be modified
			S:5/0	Math Overflow Bit	(Read Only)
B	3	Bits	B3:5/0 (B3/80)	Word 5, Bit Zero (Bit 80)	
T	4	Timers	T4:2	Timer 2	
			T4:2.PRE	Timer 2 Preset	Signed Integer
			T4:2.ACC	Timer 2 Accumulator	Signed Integer
			T4:2/DN	Timer 2 Done	
			T4:2/TT	Timer 2 Timing	
C	5	Counters	C5:8	Counter 8	
			C5:8.PRE	Counter 8 Preset	Signed Integer
			C5:8.ACC	Counter 8 Accumulator	Signed Integer
			C5:8/DN	Counter 8 Done Bit	
R	6	Control	R6:1	Control File 1	
			R6:1.LEN	Control File 1 Length	
			R6:1.POS	Control File 1 Position	
			R6:1/ER	Control File 1 Error	
N	7	Integers	N7:20	Integer Word 20	Signed Integer
			N7:20/6	Integer Word 20, Bit 6	
F	8	Float	F8:5	Floating Point Value Number 5	REAL Data type, 32 bits
ST	User	String	ST9:1	String 1 (File 9)	
A		ASCII			

For detailed information, see Allen-Bradley Instruction Set Reference Manual 1747-rm001_-en-p

Files above F8 are defined by the Programmer, to a maximum of 255 files.

SLC and MicroLogix Instructions

Basic Instructions

Mnemonic	Name	Purpose
XIC	Examine If Closed, Examine On	Examines a bit for an ON condition
XIO	Examine If Open, Examine Off	Examines a bit for an OFF condition
OTE	Output Energize	Turns a bit ON or OFF
OTL	Output Latch	Sets a bit ON when executed, the bit retains its state until unlatched or the register is cleared
OTU	Output Unlatch	Resets a bit OFF when executed
OSR	One-Shot Rising	Triggers a one time event on leading edge of signal, ON for one scan
OSF	One-Shot Falling	Triggers a one time event on falling edge of signal, ON for one scan
ONS	One-Shot (Rising)	Same as OSR
TON	Timer On-Delay	Counts timebase intervals when the energizing instruction is TRUE
TOF	Timer Off-Delay	Counts timebase intervals when the energizing instruction is FALSE
RTO	Retentive Timer	Counts timebase intervals when the energizing instruction is TRUE, retains accumulated value when energizing instruction is FALSE
CTU	Count Up	Increments the accumulated value at each false to true transition and retains the accumulated value when the instruction goes false or when power cycle occurs
CTD	Count Down	Decrements the accumulated value at each false to true transition and retains the accumulated value when the instruction goes false or when power cycle occurs
HSC	High-speed Counter	Counts high-speed pulses from a fixed controller high-speed input
RES	Reset	Resets the accumulator value and status bits of a timer or counter. *Do not use with TOF Timers!

Comparison Instructions

Mnemonic	Name	Purpose
EQU	Equal	Test whether two values are equal
NEQ	Not Equal	Test whether two values are not equal
LES	Less Than	Test whether one value is less than another value
LEQ	Less Than or Equal	Test whether one value is less than or equal to another value
GRT	Greater Than	Test whether one value is greater than another value
GEQ	Greater Than or Equal	Test whether one value is greater than or equal to another value
MEQ	Masked Equal	Test portions of two values to determine whether they are equal through a mask
LIM	Limit Test	Test whether one value is within the range of two other values

Math Instructions

Mnemonic	Name	Purpose
ADD	Add	Adds source A to source B and stores the result in the destination
SUB	Subtract	Subtracts source B from source A and places the result in the destination
MUL	Multiply	Multiplies source A by source B and places the result in the destination
DIV	Divide	Divides source A by source B and places the result in the destination
DDV	Double Divide	Divides the contents of the math register by the source and stores the result in the destination and the math register
CLR	Clear	Sets all bits of a word to zero
SQR	Square Root	Calculates the square root of the source and places the result in the destination
SCP	Scale with Parameters	Produces a scaled output value that has a linear relationship between the input and scaled values
SCL	Scale Data	Multiplies the source by a specific rate, adds to an oddest value, and stores the result in the destination
RMP	Ramp	Provides the ability to create linear acceleration, deceleration and "S" curve ramp output data waveforms
ABS	Absolute	Calculates the absolute (positive) value of the source and places the result in the destination
CPT	Compute	Evaluates an expression and places the result in the destination
SWP	Swap	Swaps the low and high bytes of a specified number of words in a bit, integer, ASCII or string file
COS	Cosine	Takes the cosine of a number and stores the result in the destination
SIN	Sine	Takes the sine of a number and stores the result in the destination
TAN	Tangent	Takes the tangent of a number and stores the result in the destination

ASN	Arc Sine	Takes the arc sine of a number and stores the result (in radians) in the destination
ACS	Arc Cosine	Takes the arc cosine of a number and stores the result (in radians) in the destination
ATN	Arc Tangent	Takes the arc tangent of a number and stores the result (in radians) in the destination
LN	Natural Log	Takes the natural log of the value in the source and stores the result in the destination
LOG	Log to the base 10	Takes the log base 10 of the value in the source and stores the result in the destination

Data Handling Instructions

Mnemonic	Name	Purpose
TOD	Convert to BCD	Converts the integer source value to BCD format and stores it in the destination
FRD	Convert from BCD	Converts the BCD source value to an integer and stores it in the destination
DEG	Convert from Radians to Degrees	Converts radians (source) to degrees and stores the result in the destination
RAD	Convert from Degrees to Radians	Converts degrees (source) to radians and stores the result in the destination
DCD	Decode 4 to 1 of 16	Decodes a 4-bit value (0 to 15), turning on the corresponding bit in the 16 bit destination
ENC	Encode 1 of 16 to 4	Encodes a 16 bit source to a 4 bit value. Searches the source from the lowest to the highest bit and looks at the first set bit. The corresponding bit position is written to the destination as an integer.
COP	Copy File	Copies data from the source file to the destination file
FLL	Fill File	Loads a source value into each position in the destination file
MOV	Move	Copies the source value to the destination
MVM	Masked Move	Copies the source value to part of the destination
AND	And	Performs a bitwise AND operation
OR	Or	Performs a bitwise OR operation
Mnemonic	Name	Purpose
XOR	Exclusive Or	Performs a bitwise inclusive OR operation

NOT	Not	Performs a NOT (invert) operation
NEG	Negate	Changes the sign of the source and stores it in the destination
FFL	FIFO Load	Loads a word into a FIFO (First In - First Out) stack on each false to true transition. The first word loaded is the first to be unloaded.
FFU	FIFO Unload	Unloads a word from a FIFO (First In - First Out) stack on each false to true transition. The first word loaded is the first to be unloaded.
LFL	LIFO Load	Loads a word into a LIFO (Last In - First Out) stack on each false to true transition. The last word loaded is the first to be unloaded.
LFU	LIFO Unload	Unloads a word from a LIFO (Last In - First Out) stack on each false to true transition. The last word loaded is the first to be unloaded.

Program Flow Instructions

Mnemonic	Name	Purpose
JMP, LBL	Jump to Label and Label	Jump forward or backward to a specified "Label" instruction
JSR	Jump to Subroutine	Jump to a designated subroutine or ladder
SBR	Subroutine label	Designates the start of a subroutine or ladder
RET	Return from subroutine	Returns from a subroutine to the point from which it was called
MCR	Master Control Reset	Turn off all non-retentive outputs in a section of ladder
TND	Temporary End	Mark a temporary end that halts program execution
SUS	Suspend	Identifies specific conditions for program debugging and system troubleshooting
IIM	Immediate Input with Mask	Program an immediate input update using a mask
IOM	Immediate Output with Mask	Program an immediate output update using a mask
REF	Refresh	Interrupt the program scan to update the I/O and service communications

Application Specific Instructions

Mnemonic	Name	Purpose
BSL, BSR	Bit Shift Left or Right	Loads a bit of data into a bit array, shifts the pattern through the array and unloads the last bit of data in the array. BSL shifts data to the left, while BSR shifts it to the right
SQO, SQC	Sequencer Output and Sequencer Compare	Controls sequential machine operations by transferring 16 bit data through a mask to image addresses
SQL	Sequencer Load	Captures referenced conditions by manually stepping the machine through its operating sequences
TDF	Compute Time Difference	Calculates the number of 10 microsecond ticks between any two captured time stamps
FBC	File Bit Compare	Compares bits between two different files
DDT	Diagnostic Detect	Used to monitor machine or process operations to detect malfunctions
RPC	Read Program Checksum	Copies the program checksum from processor memory or from the memory module into the data table

ASCII Instructions

Mnemonic	Name	Purpose
ABL	Test ASCII Buffer for Line	Determine the number of characters in the buffer up to and including the user configured end of line characters
ACB	Number of ASCII Characters in Buffer	Determine the total number of characters in the buffer
ACI	String to Integer	Convert a string to an integer value
ACL	ASCII Clear Buffer	Clear the receive and /or transmit buffers
ACN	String Concatenate	Combine two strings into one
AEX	String Extract	Extract a portion of a sring to create a new string
AHL	ASCII Handshake Lines	Set or reset modem handshake lines
AIC	Integer to String	Convert an integer value into a string
ARD	ASCII Read Characters	Read characters from the input buffer and place them into a string
ARL	ASCII Read Line	Read one line of characters from the input buffer and place them into a string

ASC	String Search	Search a string
ASR	ASCII String Compare	Compare two strings
AWA	ASCII Write with Append	Write a string with user configured characters appended
AWT	ASCII Write	Write a string

Block Transfer and PID Instructions

Mnemonic	Name	Purpose
BTR	Block Transfer Read	Receive data from a remote device
BTW	Block Transfer Write	Send data to a remote device
PID	Proportional Integral Derivative	Controls physical properties such as temperature, pressure, liquid level or flow rate using closed process loops

Interrupt Routine Instructions

Mnemonic	Name	Purpose
	User Fault Routine	Provides the option of preventing a processor shutdown
STI	Selectable Timed Interrupt	Allows you to interrupt the scan of the main program file automatically, on a periodic basis, to scan a specified subroutine file
STD	Selectable Timed Disable	Disables STIs from occurring
STE	Selectable Timed Enable	Enables STIs to occur
STS	Selectable Timed Start	Sets or changes the file number or setpoint frequency of the STI routine
DII	Discrete Input Interrupt	Allows the processor to execute a subroutine when the input pattern of a discrete input card matches a compare value that you programmed.
ISR	I/O Interrupt	Allows a specialty I/O module to interrupt the normal processor operating cycle in order to scan a specific subroutine file
IID	I/O Interrupt Disable	Disables I/O Interrupts from occurring
IIE	I/O Interrupt Enable	Enables I/O Interrupts to occur
RPI	Reset Pending Interrupt	Aborts a pending I/O Interrupt
INT	Interrupt Subroutine	Optional instruction to identify interrupt subroutines

Communication Instructions

Mnemonic	Name	Purpose
SVC	Service Communications	Interrupts the program scan to execute the service communication portion of the operating cycle
MSG	Message Read/Write	Transmits data from one node to another on the network
CEM	ControlNet Explicit Message	Transmits CIP generic commands to other ControlNet nodes via the 1747-SCNR
DEM	DeviceNet Explicit Message	Transmits CIP generic commands to other DeviceNet nodes via the 1747-SDN
EEM	EtherNet/IP Explicit Message	Transmits CIP generic commands to other Ethernet/IP nodes via channel 1

For detailed information, see Allen-Bradley Instruction Set Reference Manual 1747-rm001_-en-p

Starting and Editing a Project with RSLogix 500

Allen Bradley's RSLogix 500 software is installed in the Rockwell Software folder under RSLogix 500 English. It may also be on your desktop as a shortcut, as shown in the icon on the left. The latest version as of June 2017 is version 11.0.

The extension for RSLogix 500 programs is .RSS, double-clicking on a file will also open the programming software. When the software is first opened, it will look as shown in the image below:

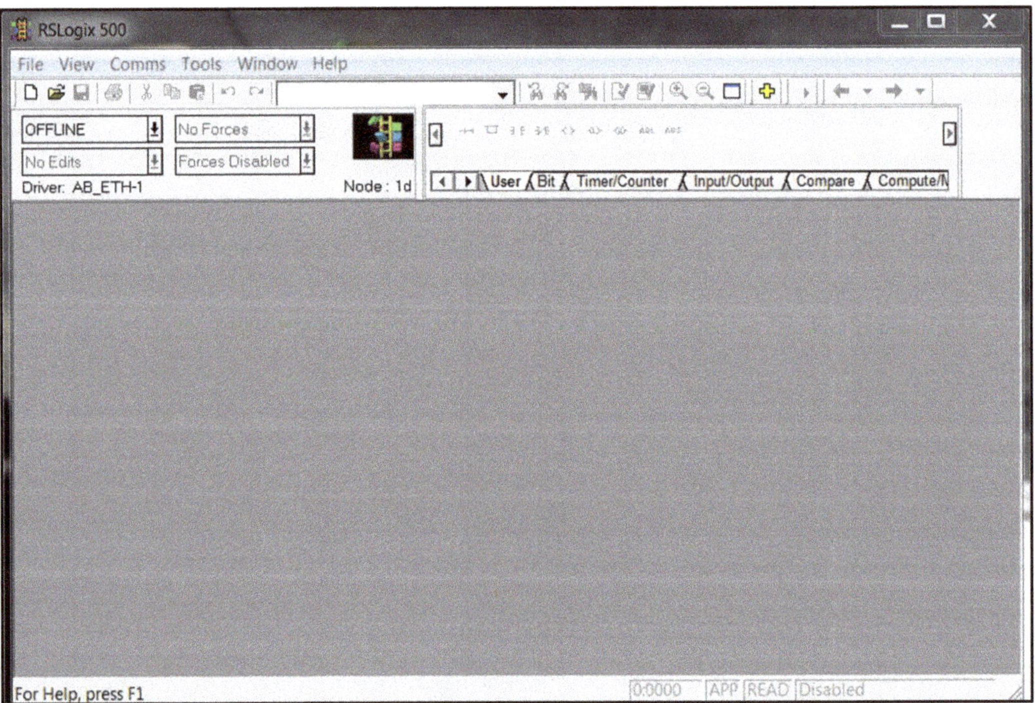

Creating a Project

To create a new project, select "New" under the File tab. The first screen that will appear will ask you to choose and name your processor. The processor name can only be a maximum of 8 characters; a typical name might contain an abbreviation or number for a machine or system, such as "PNTLINE" (Paint Line), "RLMILL42" (Roll Mill 42) or the like.

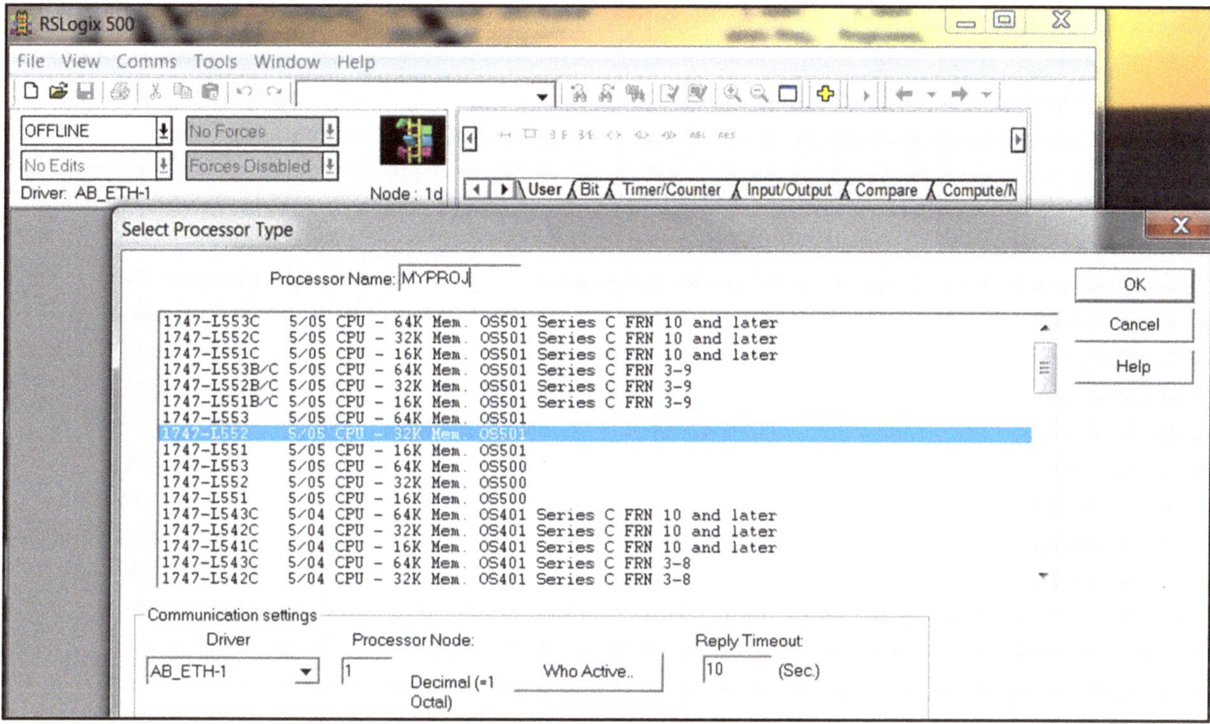

A communication settings dialog is also shown on this screen; don't worry about setting it up yet. This will be done later using the RSLinx communications software.

Until the hardware platform is selected, the software doesn't know how to set up the data tables and properties. It is important to note that the Firmware Revision Number (FRN) selection is part of the hardware selection process.

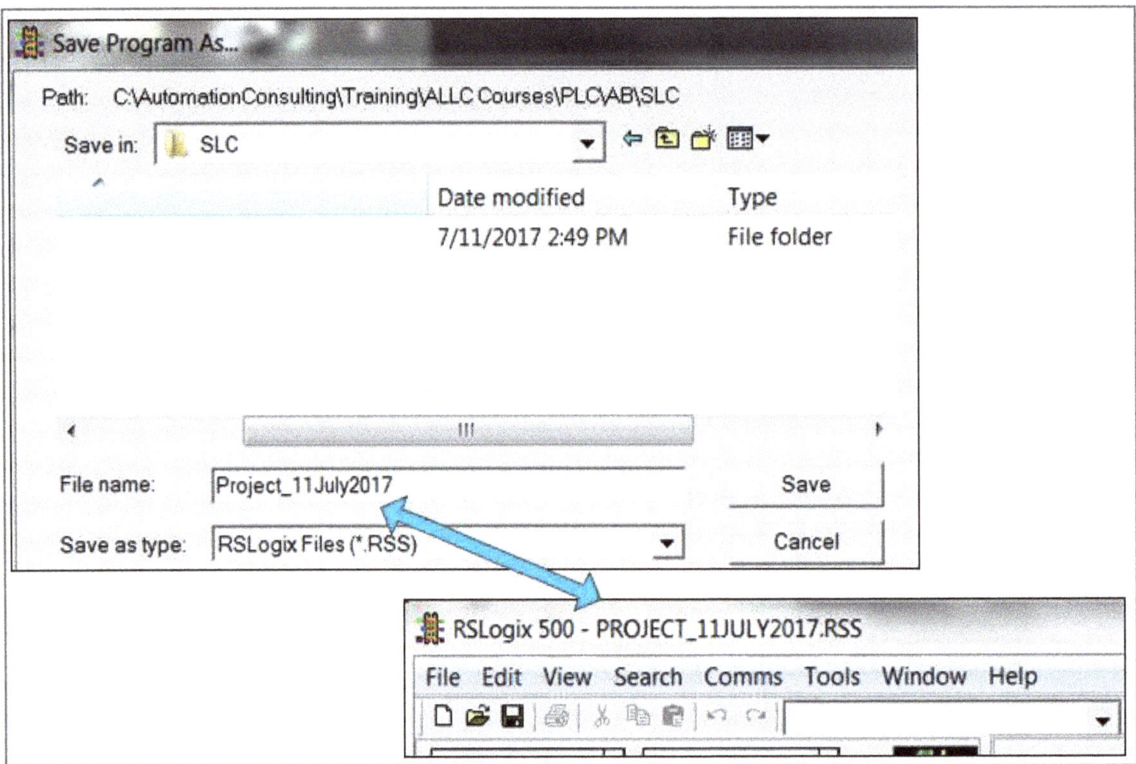

Initially, the project will be saved under the name of the processor. Selecting "Save As" will allow you to save with a date and a different title.

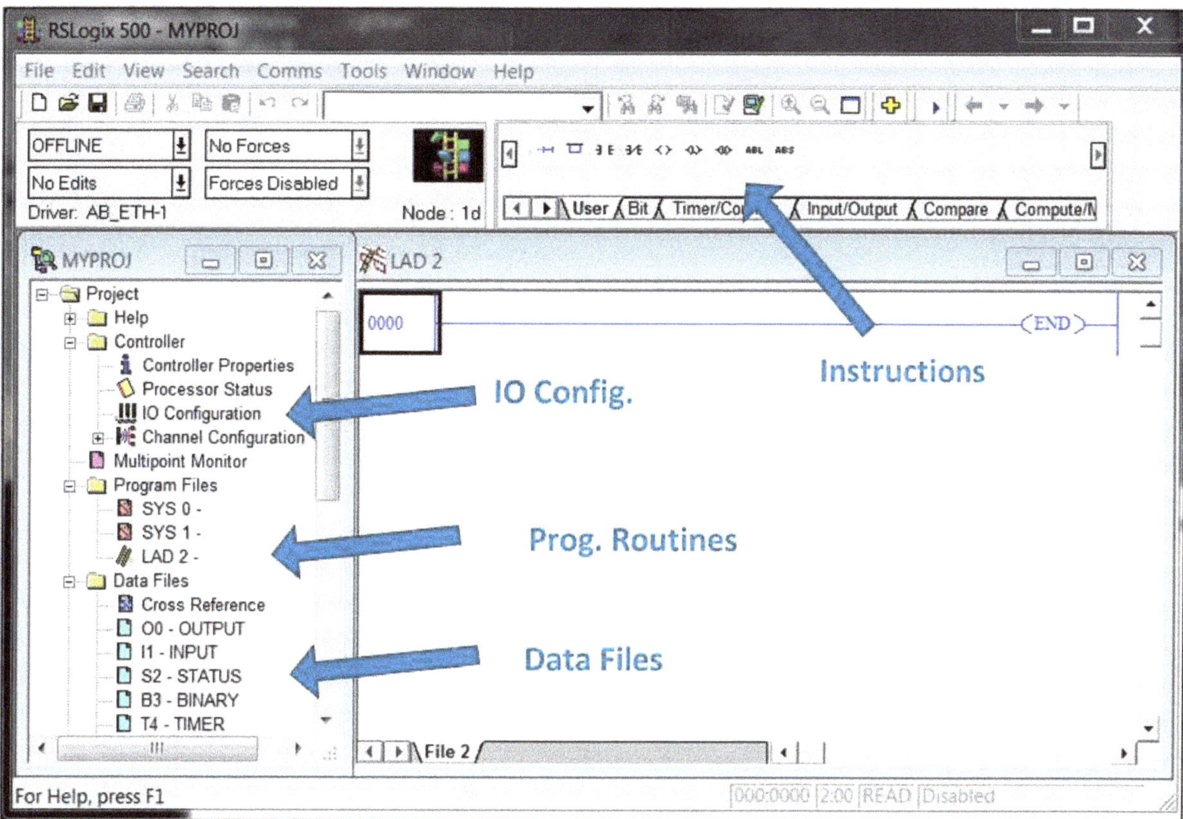

After creating the project, the programming environment will appear as shown above. The area on the left is where most of the items you will be adding to the project are located; the first task you will want to accomplish is to configure the I/O.

Hardware Configuration

Double-clicking the I/O Configuration icon will open a selection window containing all the different types of I/O cards available. This is also where you select your rack size.

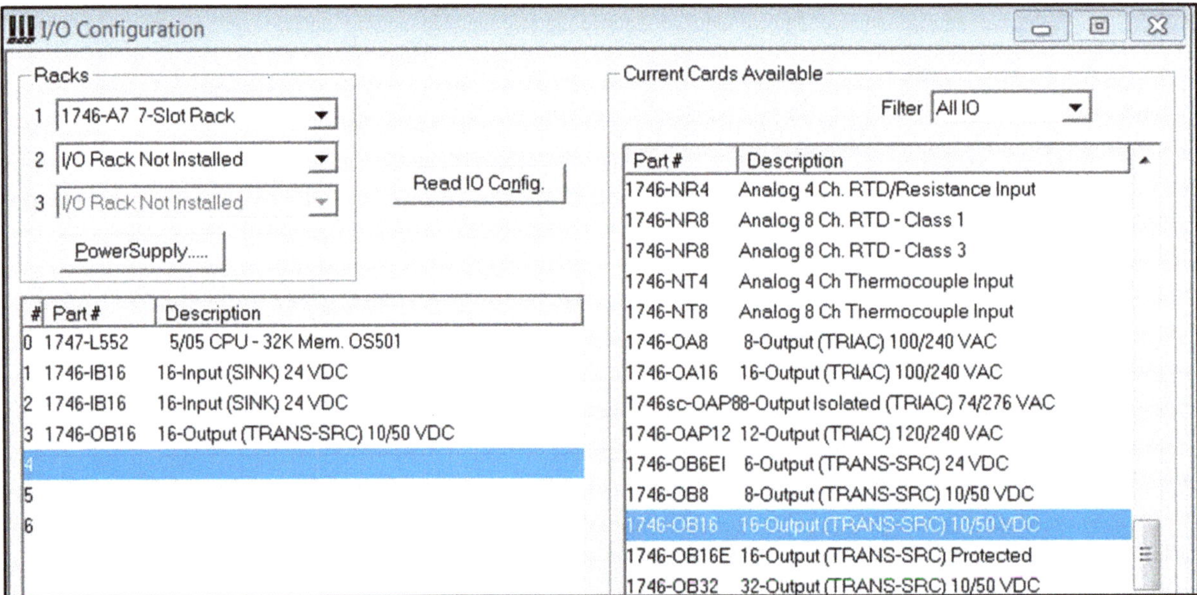

SLC processors are always located in slot 0 on the left side of the rack next to the power supply. Notice the power supply button; this must also be selected.

MicroLogix projects are configured similarly, but some may have I/O integrated with the processor. Also, some of the MicroLogix PLCs have a "rackless" design that allow I/O cards to plug in from the side, building the backplane as you go.

Writing the Program

After configuring the hardware, programming can begin. Organization of the code is done within Ladder 2, which is designated to run first. When the program is first created, LAD 2 (the main routine) is automatically created. Other subroutines are created by right-clicking the Program Files folder and selecting "New".

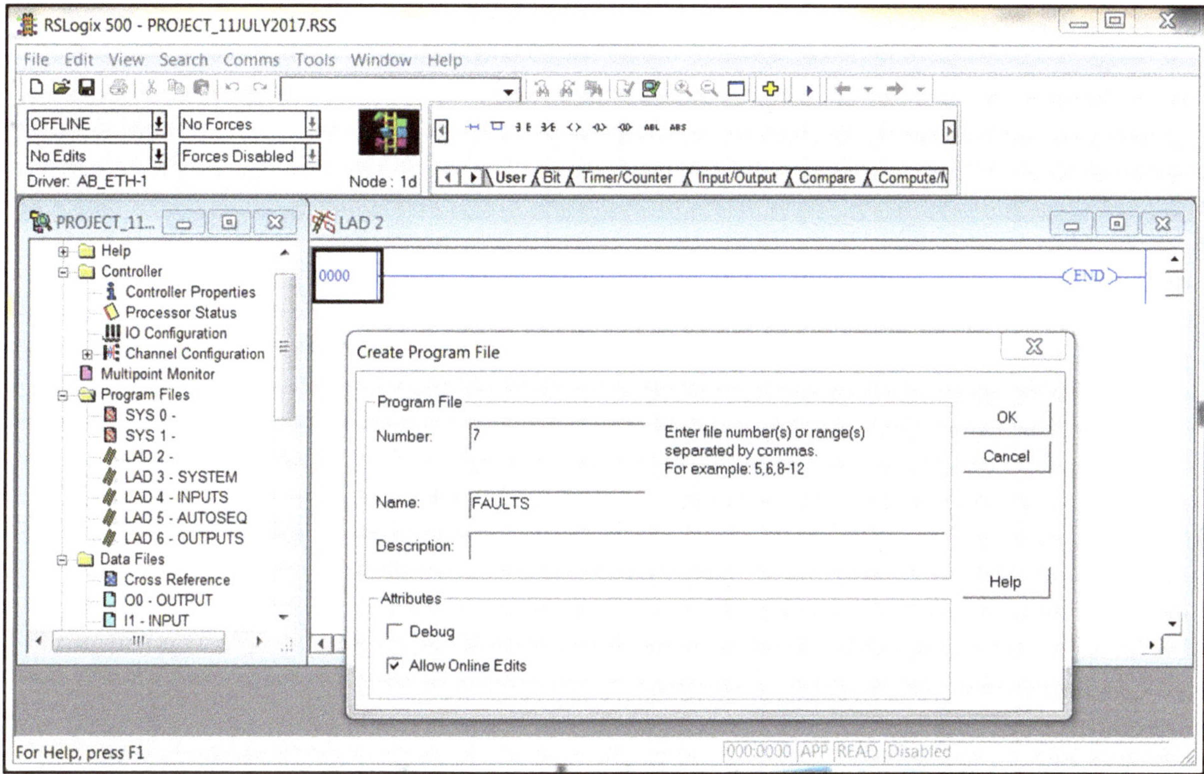

Routines are numbered from 2 to a maximum of 255. The only routine that runs automatically is LAD 2; all other routines need to be called using a **JSR** (*Jump to Subroutine*) instruction.

If you don't call the subroutine, the code inside it will not run!

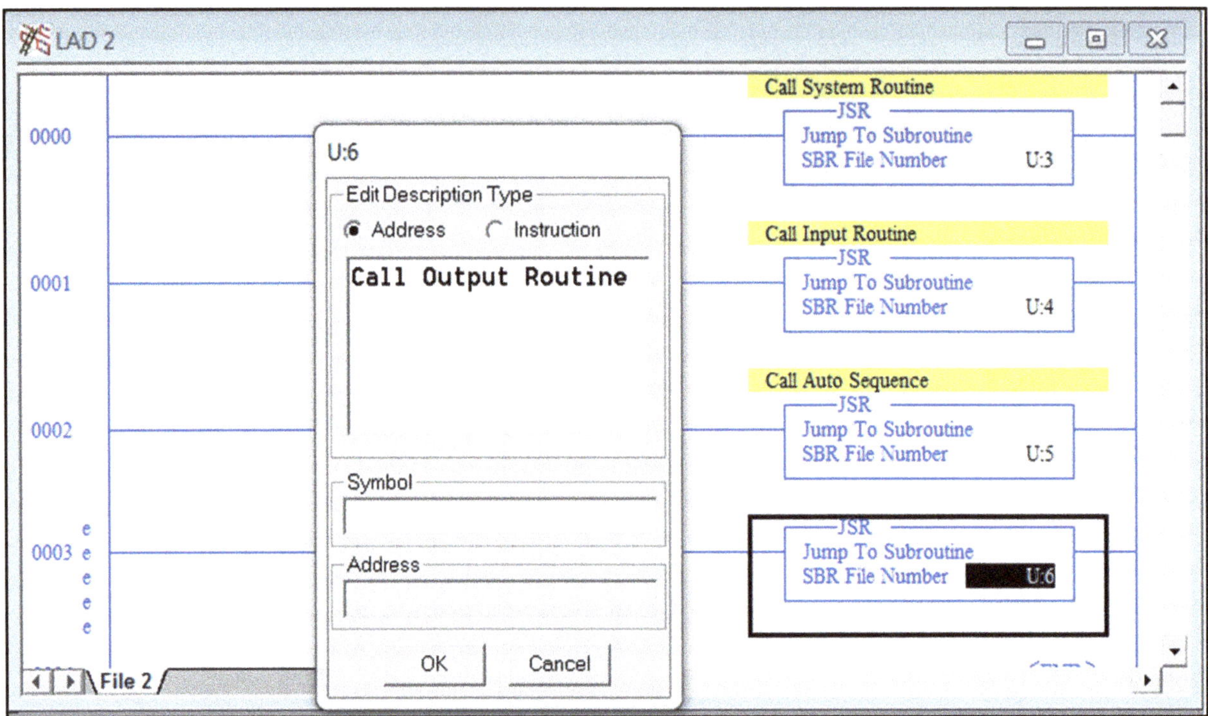

It is typical to use LAD 2 almost exclusively for subroutine calls. Note that the rungs don't have a contact in front of the JSR instruction; they are called unconditionally. This helps in organization, separating different functions of the program into sections so that they can be easily followed.

The lower case "e" on rung 0003 indicates that it is being edited. Instructions can be selected from the tabs in the "Instructions" section or a mnemonic can be typed at the beginning of the rung. In this case, "JSR" was typed with the cursor over the 0003 label and the dialog box for the instruction appeared as shown.

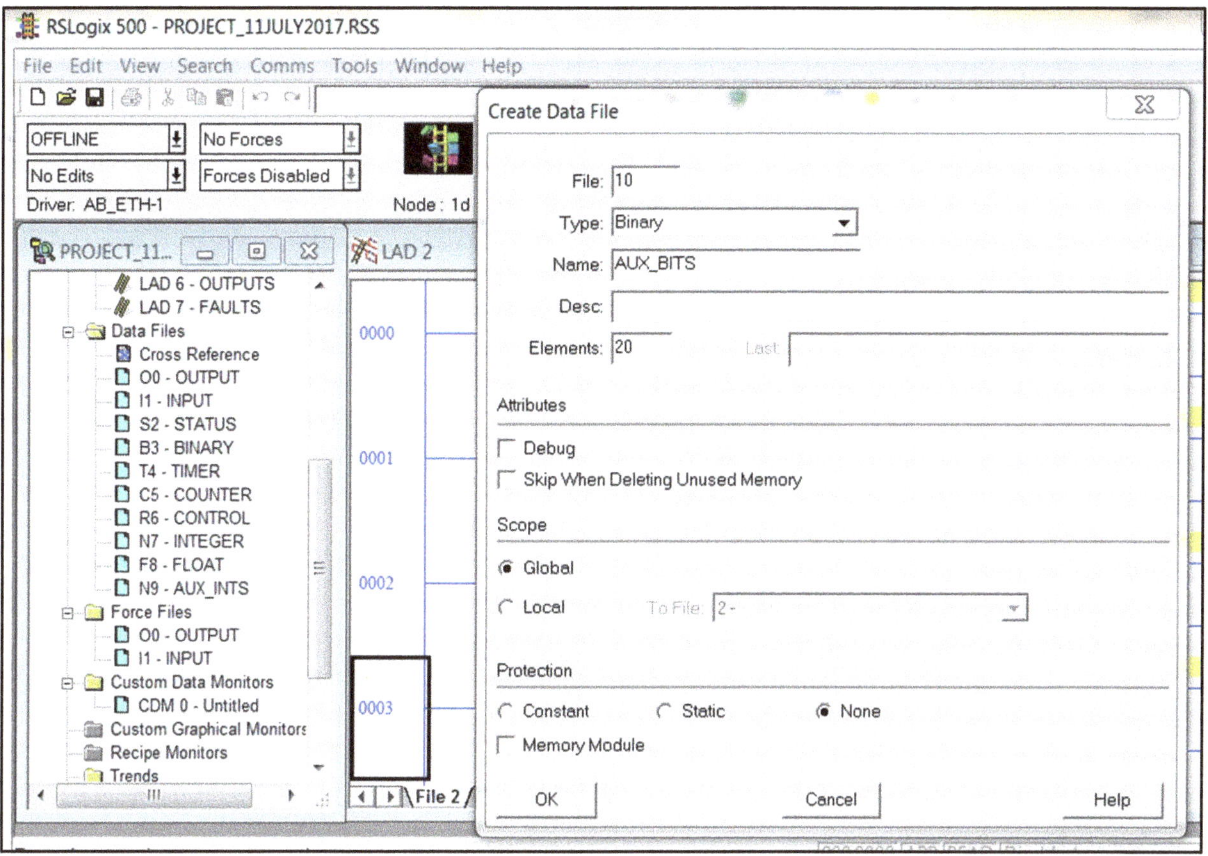

Data files can be added by right-clicking the Data Files folder and selecting "New". As described in the next section on Memory Registers, different types of data registers can be added as required. The Input and Output (O0 and I1) registers are created automatically based on hardware selection, while B3-F8 are created with the program. An important note on these registers: they only contain one element each when the program is created; you will need to go into the registers and edit the "Elements" field to size the registers. A maximum of 255 elements can be designated, but memory is subject to the limits of the processor chosen.

Discrete Instructions

Bit-level instructions are referred to by the mnemonics in the Basic Instructions list earlier in this book. *Normally open* (**NO**) contacts are listed as **XIC** (e**X**amine **I**f **C**losed), and *normally closed* (**NC**) contacts are listed as **XIO** (e**X**amine **I**f **O**pen). These and all other bit instructions can be found in the Bit tab at the top of the RSLogix 500 software. Coils are designated as **OTE** (**O**u**T**put **E**nergize) for a momentary coil, **OTL** for a *latch* or set bit, and **OTU** for an *unlatch* or reset.

One shots are designated as **ONS**, **OSR** (rising/leading edge) or **OSF** (falling/trailing edge). These can be dragged from the toolbar or inserted after an object on a rung by double-clicking an instruction on the toolbar while a contact is highlighted on the rung.

Timers and Counters

Allen-Bradley 500 platform timers are integer based; that is, the Preset and Accumulator values are in integer form. This means that a time base is required, which can be .001, .01 or 1 second. For the Cycle Start Timer shown here, 300 x 0.01 means that the setpoint for the timer is 3 seconds, after which the .DN (Timer Done) bit will be true.

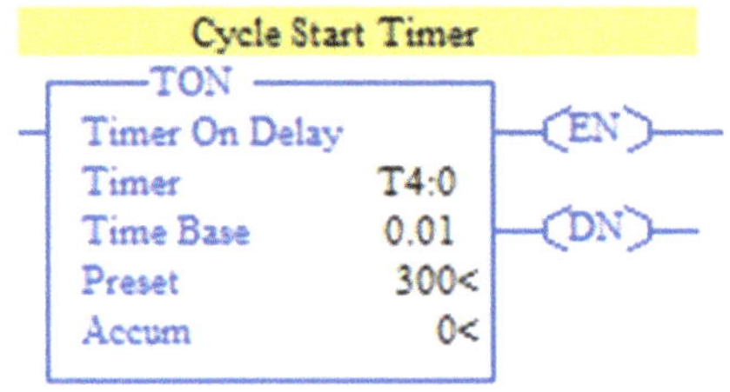

In addition to the standard Preset and Accumulated values and the Done bit, an Enable (.EN) and Timer Timing (.TT) bit is available for use. These are accessed by placing a contact on the rung and typing in the address of the timer (T4:0 in the above example) followed by the element. If the element is a BOOL, a forward slash will be used (T4:0/DN, T4:0/TT and T4:0/EN). If the Preset or Accumulated value is accessed, a dot or period is used (T4:0.PRE or T4:0.ACC).

On-Delay (**TON**), *Off-Delay* (**TOF**) and *Retentive On-Delay* (**RTO**) timers are available for Allen-Bradley platforms. If it is desired to have a new timer file, any unused file number 9-255 can be used to create a new group. This is explained further in the Files section of this book.

To view the data structure and status of timers or counters the data file can be opened by double-clicking, or right-clicking the file and selecting Open.

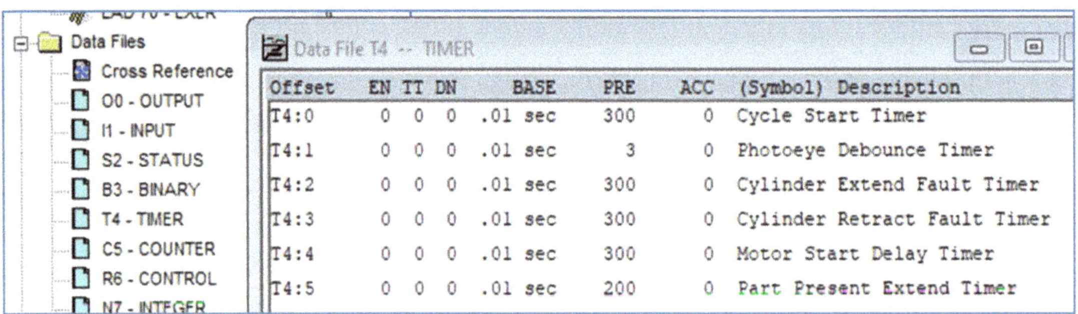

Data File T4 -- TIMER						
Offset	EN TT DN	BASE	PRE	ACC	(Symbol) Description	
T4:0	0 0 0	.01 sec	300	0	Cycle Start Timer	
T4:1	0 0 0	.01 sec	3	0	Photoeye Debounce Timer	
T4:2	0 0 0	.01 sec	300	0	Cylinder Extend Fault Timer	
T4:3	0 0 0	.01 sec	300	0	Cylinder Retract Fault Timer	
T4:4	0 0 0	.01 sec	300	0	Motor Start Delay Timer	
T4:5	0 0 0	.01 sec	200	0	Part Present Extend Timer	

Counters are accessed as up (**CTU**) or down (**CTD**) counters. As with timers, the Preset and Accumulator values are in integer form. Since integers in Allen Bradley are signed, this means that the maximum value for an RSLogix500 counter is 32,767 and the minimum value is -32,768. If the values are exceeded, the Overflow (.OV) or Underflow (.UN) bits will be active.

The counter elements consist of the Preset (.PRE) and Accumulator (.ACC) integers and six BOOLs that are bits of a control integer.

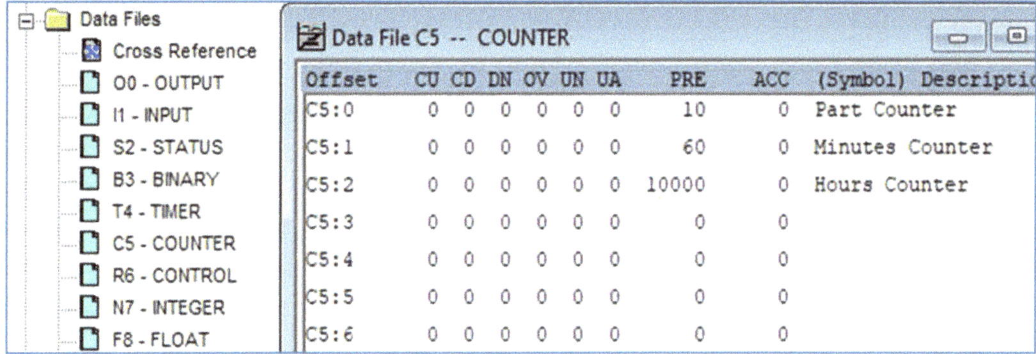

The .CU and .CD bits are active if the counter is enabled, like the .EN bit of a timer. The .DN bit is on if the Accumulator value is greater than the Preset. The .UA (Update Accumulator) bit is only used for the *High-speed Counter* (HSC), which is a hardware counter used with encoders .for motion control.

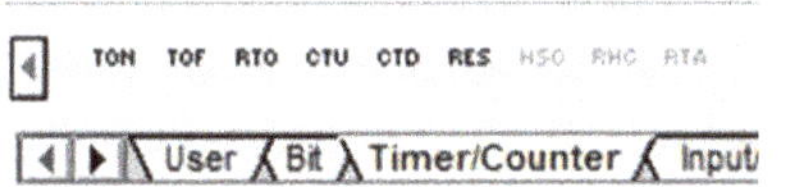

An additional element of both timers and counters is the *Reset* (**RES**) bit. This element is accessed by selecting it from the toolbar or using the RES mnemonic. It does not show in the data table. To address the bit, type the address (T4:X or C5:X) above the bit on the rung. The Reset places the Accumulator at 0 and clears the .OV and .UN bits.

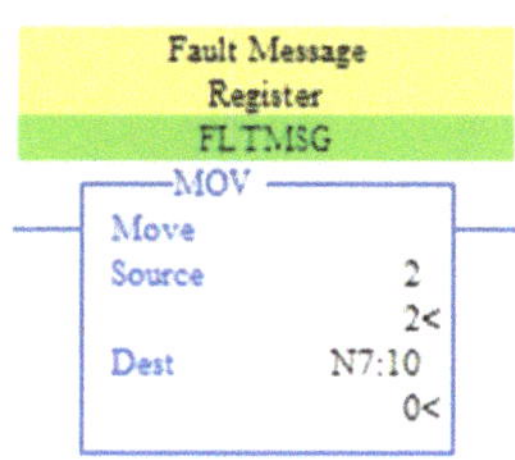

Data Manipulation and Math

Data is organized by data type into files as shown in the Data Files images above, and addressed as shown at the beginning of the **SLC and Micrologix Memory Registers** section. More will be discussed later in this section about how data files are organized.

The *Move* (**MOV**) instruction is used to copy a single number such as an integer or REAL from one location to another. There are two places where data is entered into the instruction: The **Source** which is the number to be copied, and the **Destination** which is where it will be copied to. The Source can be a constant or a variable, while the destination must be a variable or data register.

When copying multiple items, the *Copy* **(COP)** instruction is used. This instruction allows the first address of a group to be entered into the Source and Dest fields, followed by the number of elements to be copied (Length). A "#" is placed in front of the address, signifying that this is a File type operation.

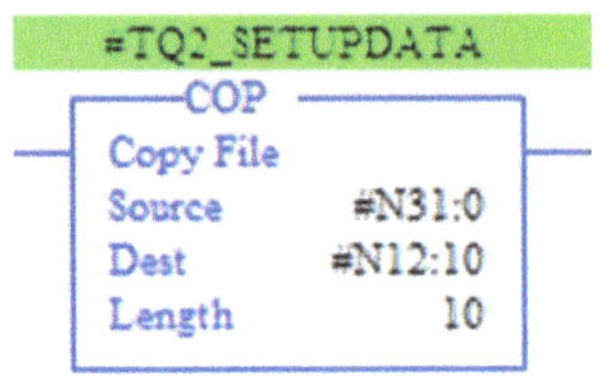

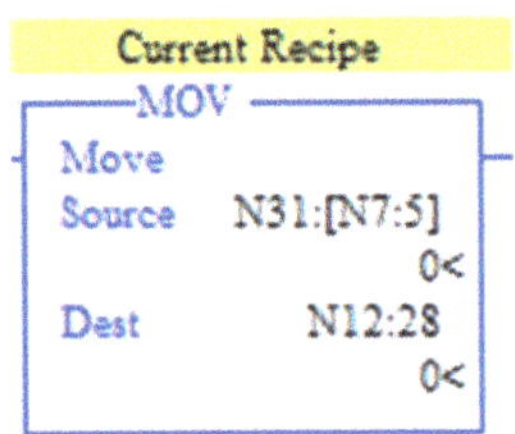

An additional useful feature when moving or copying data is the use of **Indirect Addressing**. This allows a pointer to be used to decide which element is moved. Instead of a fixed address or number, a variable can be placed inside of square brackets to change either the Source or Dest address. In the example at left, N7:5 contains the variable pointer. If for example an 8 is placed in N7:5, the number in N31:8 will be moved into N12:28. This also works with bit-level instructions, If the same pointer is used to address an output, O:4/[N7:5], and the number in N7:5 was a 2, then O:4/2 would be controlled by a rung. In the example at right, the address was entered as O:4/[N7:5] and the software converted it to the example in the picture.

Comparison instructions are found in the Compare tab and include all of the standard less than/greater than functions but also the *Limit* **(LIM)** instruction. An interesting feature of the

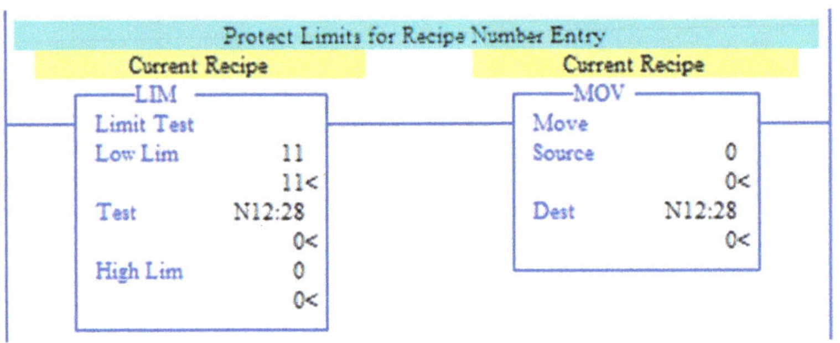

LIM is that if the higher number is placed in the Low Lim field, the instruction operates in reverse; rather than the instruction being true when the Test variable is in range, it will be true when it is outside of the limits.

Math instructions include not only the standard ADD, SUB, MUL and DIV instructions (found in the Compute/Math tab), but also *Compute* (CPT), which allows a formula to be entered, and a set of Trigonometric Functions, scaling, data conversion and more. Tabs for these instructions are in **Trig Functions** and **Advanced Math**, but it is much easier to simply type in the mnemonic. Some of the instructions in these folders also do data type conversions.

One of the most useful instructions in the RSLogix500 set is *Scale with Parameters* **(SCP)**. Raw analog inputs are in signed integer format, a typical 0-10VDC signal would be in the range 0-32,767. To convert this value into a user friendly scaled value, the raw input minimum and maximum (which may not be the full range of unsigned integers) and the desired scaled limits are entered into the instruction. The Output value can then be in Integer or REAL format.

The Min and Max limits can also be entered as variables so that they can be modified from an HMI. This is useful when calibration of instruments needs to be performed by personnel without access to the RSLogix 500 software.

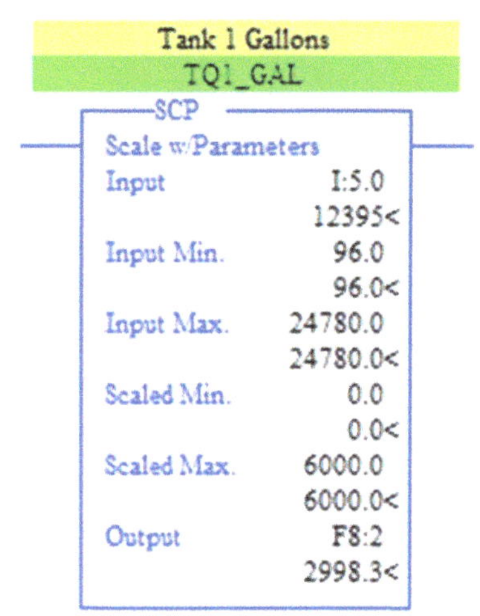

Other Instructions

There are several other tabs worth mentioning in the RSLogix 500 instruction set. The **ASCII Control** tab contains instructions to access the serial port of the CPU. There is a buffer which contains the characters received on the port, which is CH0. The Error field is a hexadecimal value that indicates *why* the ER bit is set in the control data file.

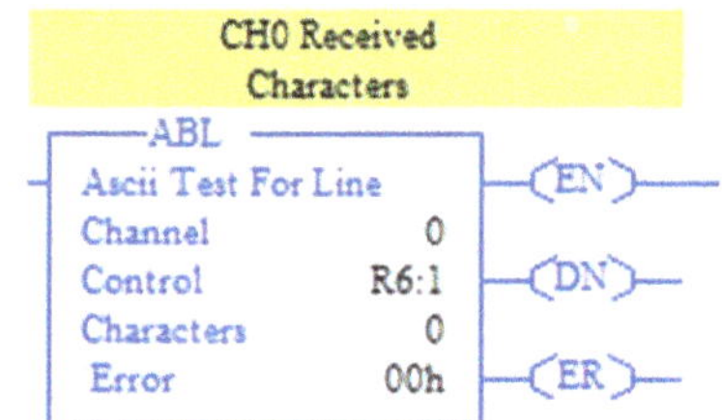

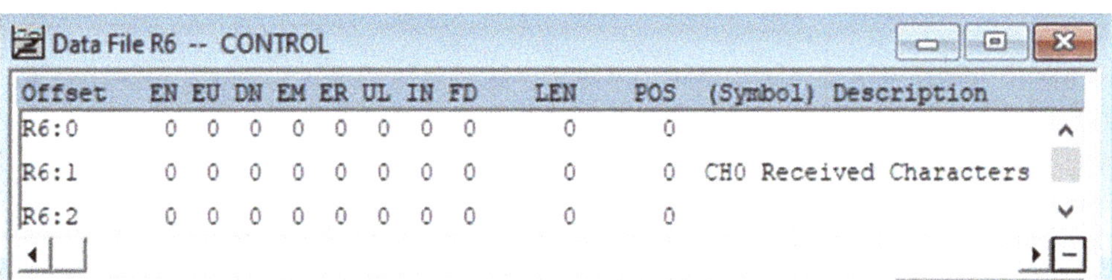

Offset	EN	EU	DN	EM	ER	UL	IN	FD	LEN	POS	(Symbol) Description
R6:0	0	0	0	0	0	0	0	0	0	0	
R6:1	0	0	0	0	0	0	0	0	0	0	CH0 Received Characters
R6:2	0	0	0	0	0	0	0	0	0	0	

Data File R6 -- CONTROL

Many of the more complex instructions for ASCII instructions, file operations, PID and others use the **R6 Data File** register shown above. This register holds bits indicating the status of its assigned instruction, and a **Length (LEN)** and **Position (POS)** integer that contain the location of searches or comparisons.

The **ASCII String** tab contains instructions that manipulate STRING tags for display or communications. Strings are arrays of characters used for human readable text or messages. The Allen-Bradley String data type also holds the length or number of characters in the register, which can be useful when locating terms within the String. RSLogix 500 Strings contain 82 characters.

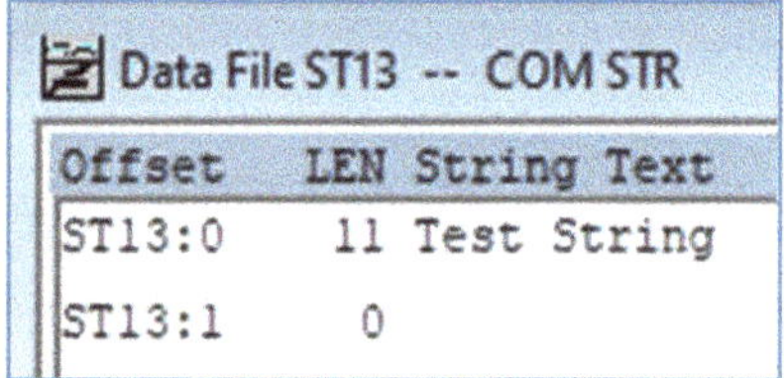

Offset	LEN	String Text
ST13:0	11	Test String
ST13:1	0	

Data File ST13 -- COM STR

A String register is not assigned automatically like the Binary, Integer and Float registers, but must be added manually.

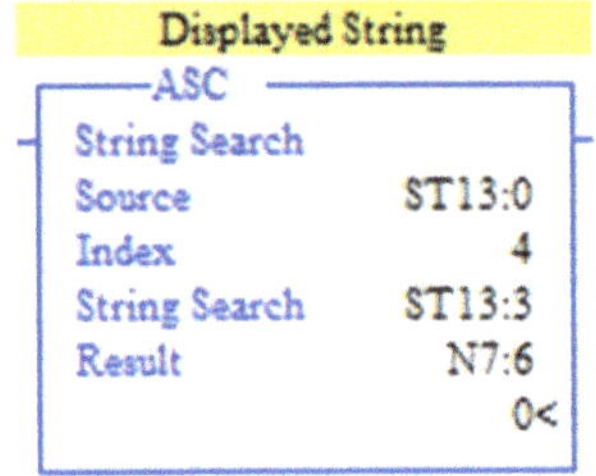

Instructions such as the **ASC** shown at left use an **Index** to show the starting point for an operation, in this case a search for another string. This may be a constant as shown, or a variable Integer file.

If an error occurs in an instruction without an R6 Data File assigned, the Status Register **S2** can be accessed by typing in the address immediately following the instruction.

S:5/2 is the address of the Control Register error. It can be used even if the R6 file isn't present in the instruction itself. It is important to use the instruction in the rung immediately after the logic being checked since the bit may be set by a different error. The Control Register Error bit can be cleared after it is used to latch a logic-specific bit.

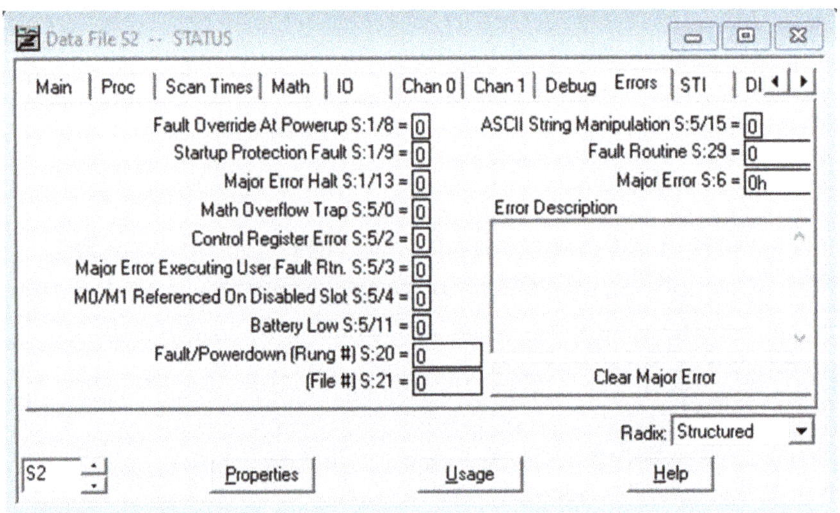

Program Control instructions allow jumping to a label in a routine, jumping to a different routine, ending a routine early, disabling output coils or analog values, or even suspending the controllers operation (**SUS**)! These are among the most potentially dangerous instructions in the RSLogix 500 platform, and they should not be used unless the programmer has examined all the possible consequences of the action.

Using a *Jump* (**JMP**) instruction with a *Label* (**LBL**) below it skips over the code between the two, while placing the label above the jump will **loop**! If this is done, it is critical to place a counter or other method of exiting the loop inside, or the loop will continue forever, creating a watchdog fault.

PLC Hardware and Programming Exercises

Following are the exercises from the generic PLC Hardware and Programming book answered using RSLogix 500 software:

Exercise 5

1. Draw ladder logic to place a machine into either Auto or Manual mode by pressing physical pushbuttons. Ensure that the machine is placed into Manual if a Fault occurs.

2. Draw logic to accomplish the following:

    a. Turn on a motor using a start button and a stop button using a hold-in contact. Both buttons should be <u>wired</u> Normally Open (NO). Include a NC Fault contact that will stop the motor.

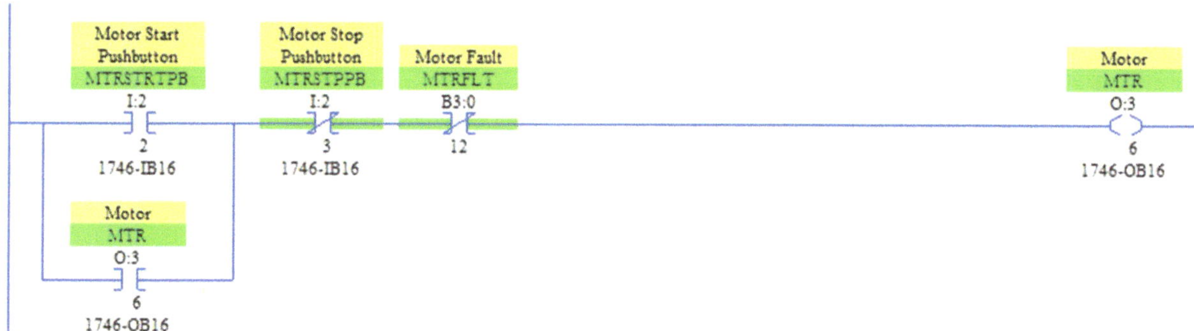

b. Latch the Fault bit if the motor's overload trips or the guard door is opened while the motor is running.

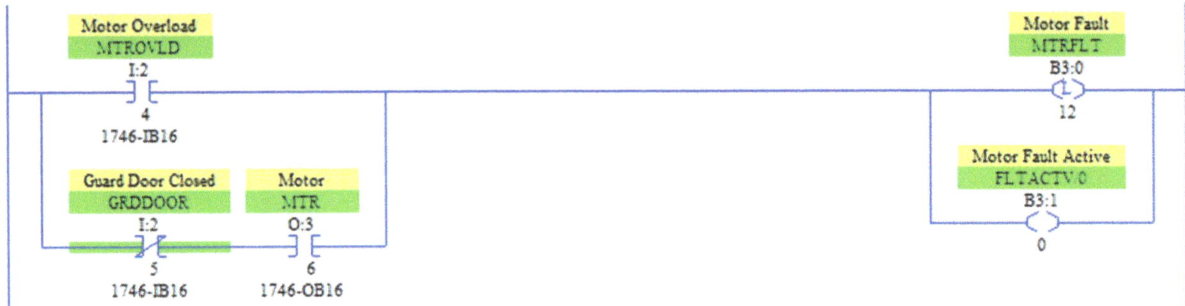

c. Turn on a Red Light if there is a fault.

d. Turn on a Green Light if the motor is running.

e. Reset the Fault if there is no fault and a Reset Button is pushed.

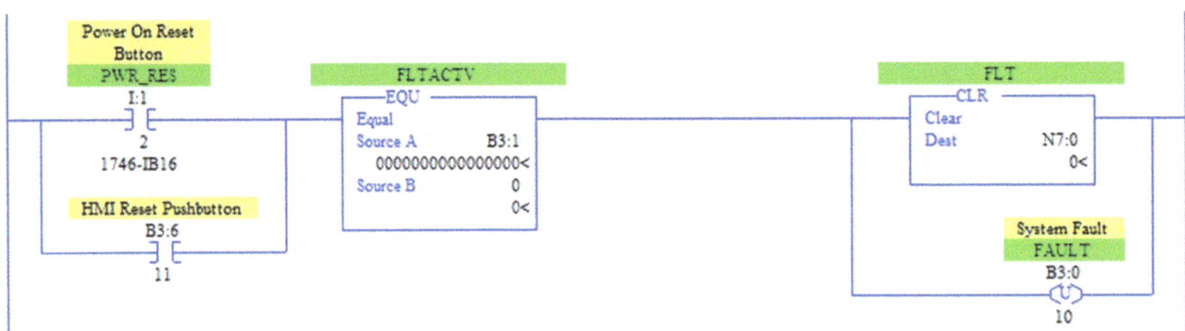

This explains why the Motor Fault Active coil was placed in the network of exercise b! The fault register can only be cleared if no fault is active.

Exercise 6

1. Draw ladder logic to delay the start of a motor for 3 seconds. Use a start and stop button to control the motor. Ensure that if the button is released before the motor starts, it will start anyway.

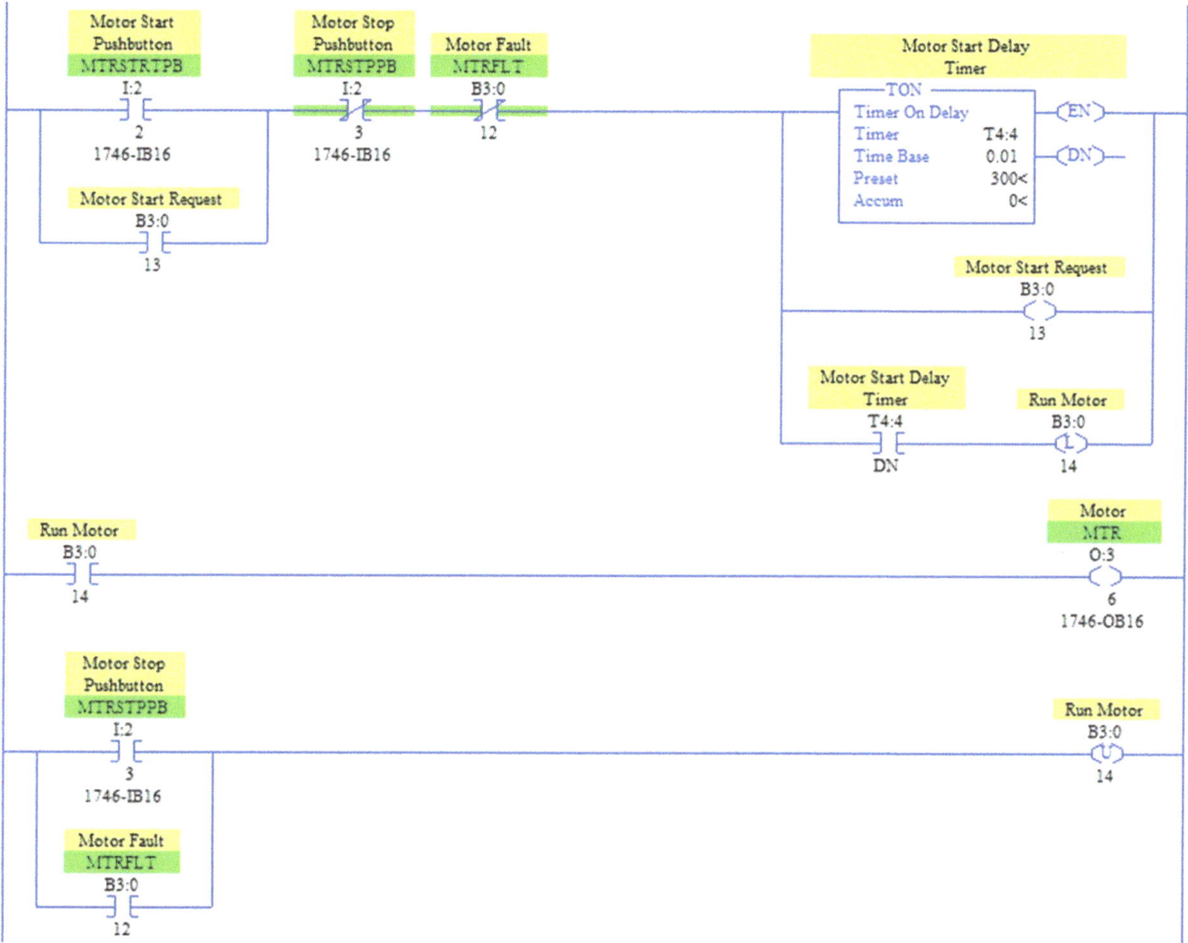

Note the Start Request hold-in contact and the integration of the motor fault.

2. Draw ladder logic to ensure that a sprayer stays on for at least 2 seconds whenever an object passes in front of a photoeye on a conveyor. Ensure that if the conveyor is not running, the sprayer stops.

Placing a one shot before the timer will make the sprayer activate for two seconds rather than adding to the time that the photoeye is on.

3. Draw ladder logic that creates a one second on, two second off train of pulses. Use a pushbutton to sound a buzzer with this signal.

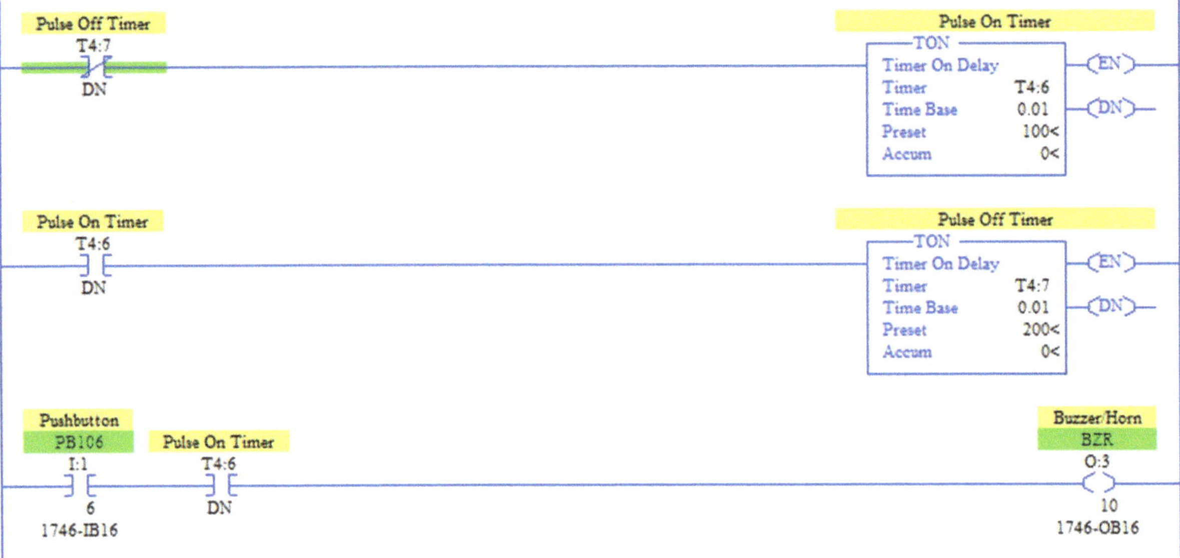

Exercise 7

1. Draw ladder logic that counts parts into a box. When the box is full (10 parts), latch a memory bit that controls a light, the light tells an operator to remove the box. Use a sensor to detect that the box has been removed. When the box has been removed, reset the counter and the memory bit that controls the light.

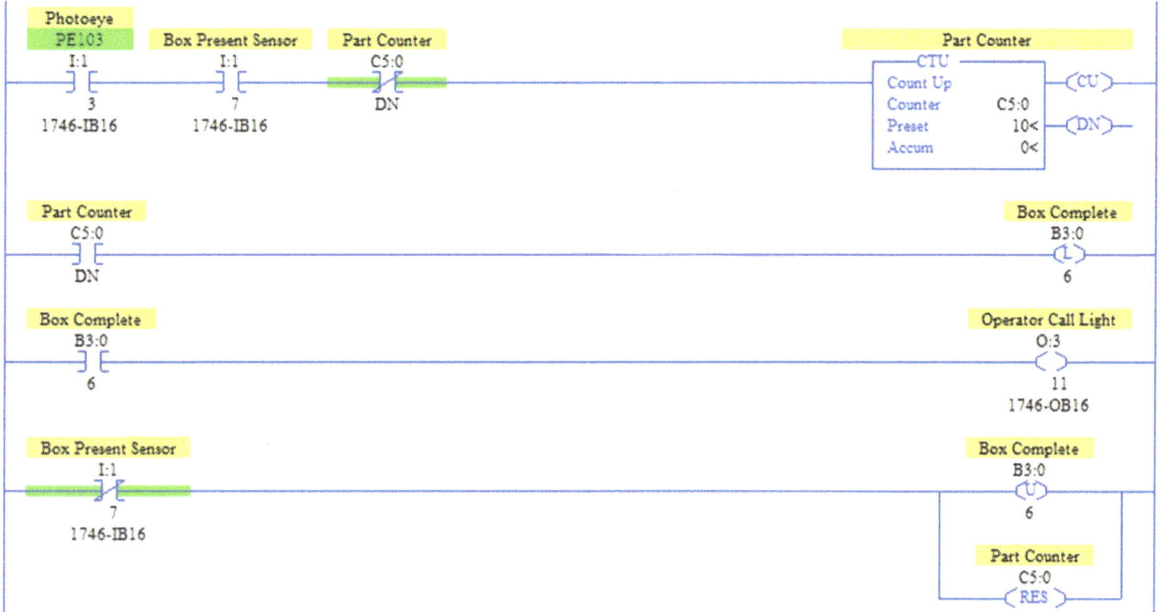

2. Draw ladder logic that uses a Retentive On-Delay timer to accumulate run time on a motor. When the timer reaches one minute, use the done bit to increment a "Minutes" counter and reset the timer. When the counter reaches 60 minutes, use its done bit to increment an "Hours" counter and reset the Minute counter. When the Hour counter reaches 10,000, it is time to perform maintenance on the motor.

   Do not reset the Hours counter automatically; use a pushbutton that is pressed by the maintenance technician, confirming that maintenance has been done. How can you incorporate logic to ensure that the technician really did do the maintenance?

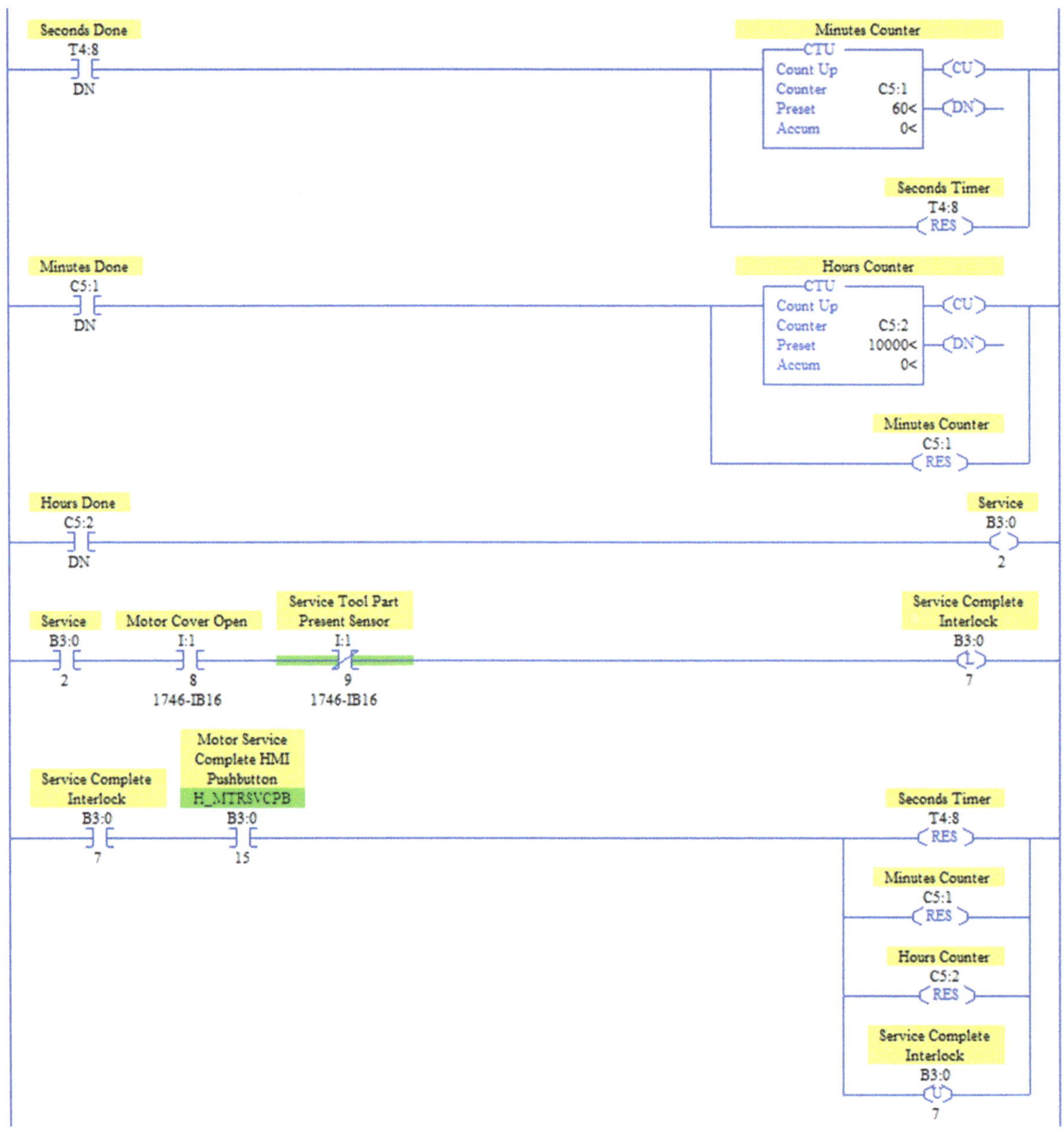

The Service Complete interlock ensures that the cover was open on the motor housing and that the service tool was removed from its location at some time after the Service bit had been activated.

Exercise 8

2. Write Ladder Logic that increments an Auto Sequence through three different steps based on I/O and/or internal bits. On the last step, reset the sequence to zero.

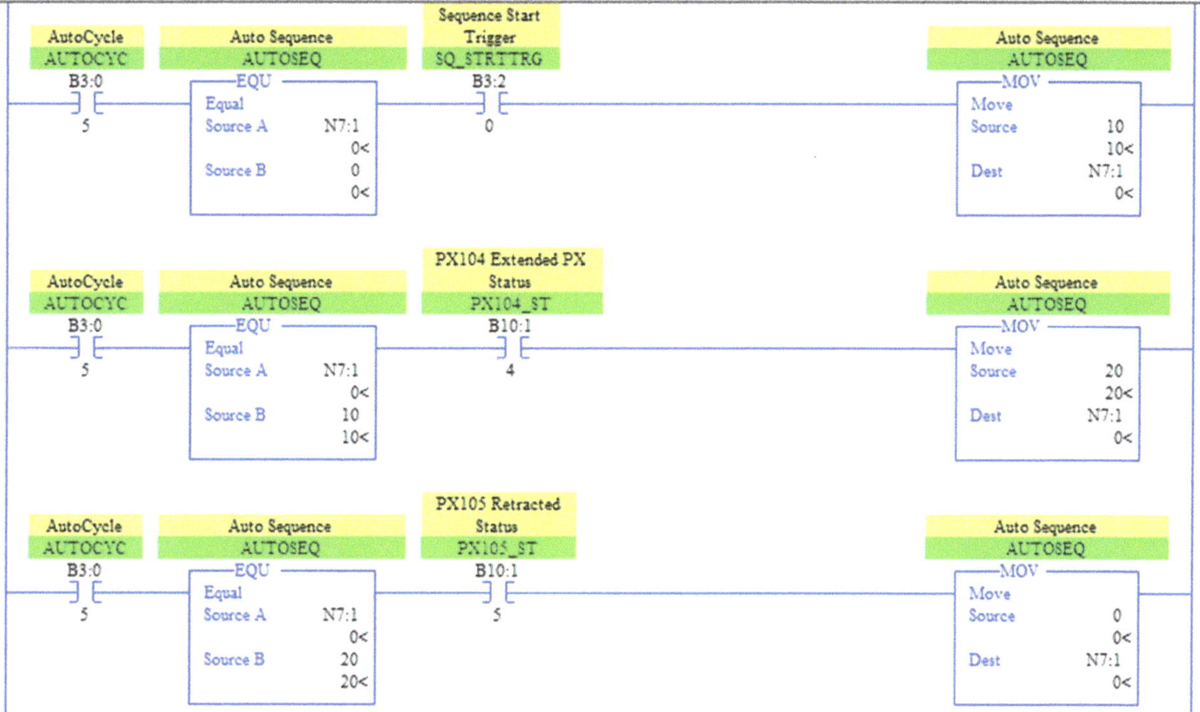

Auto Sequence steps typically increment by 10 in case an intermediate step or steps need to be inserted later.

Exercise 9

1. A Variable Frequency Drive (VFD) is used to control a conveyor. Its maximum speed is 1750 RPM, but the number sent from the drive is in integer form. At full speed, the integer reads 31,760, while it reads zero when stopped. Write ladder logic to calculate the percentage of a drive's actual speed to its total speed, also providing a REAL speed in RPM.

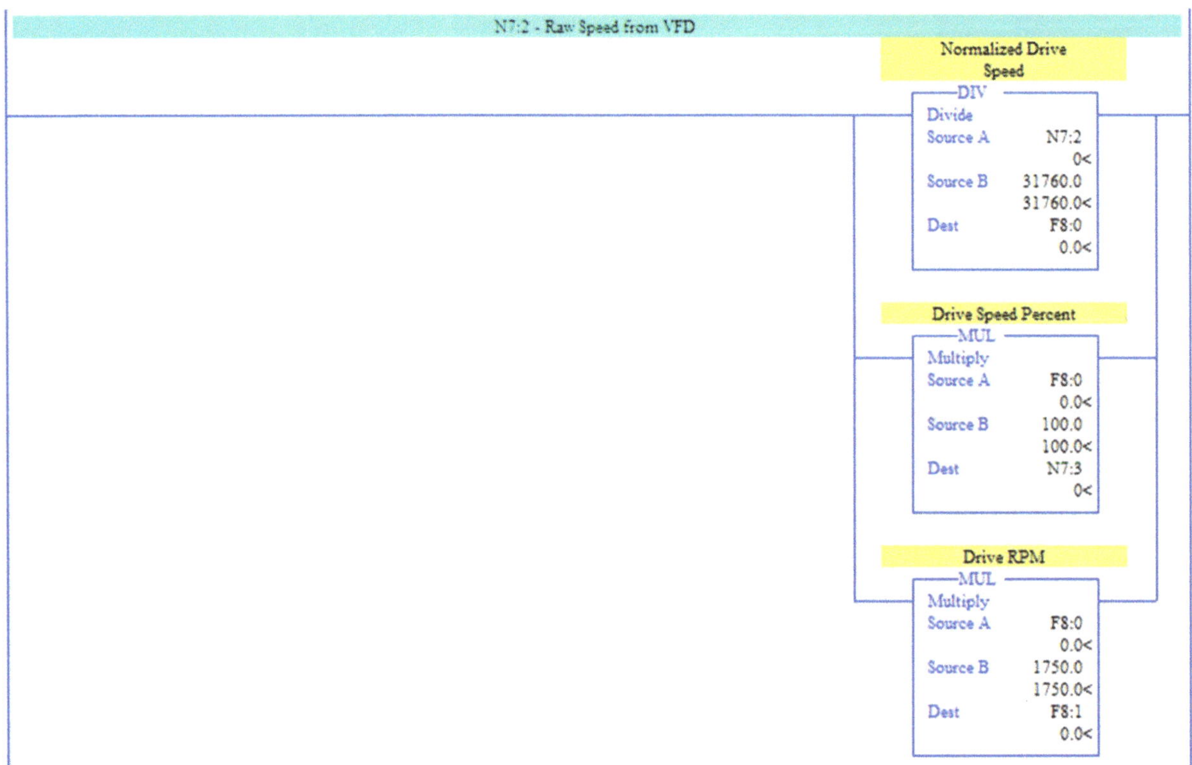

The SCP (Scale w/Parameters) instruction is also very useful for scaling and is shown below:

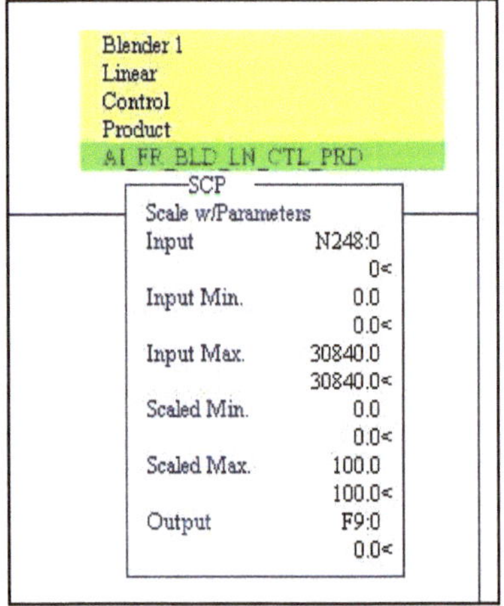

Exercise 9 Continued

2. Production and reject data from a manufacturing line is entered manually at the end of
    each shift. After completion of the third shift, there are three registers named
    Shift1_Prd, Shift2_Prd and Shift3_Prd containing the total parts made that shift, and 3
    more registers named Shift1_Rej, Shift2_Rej and Shift3_Rej that contain the number of
    failed parts. Write ladder logic to calculate Total Parts, Total Rejects and Total Good
    Parts for the day.

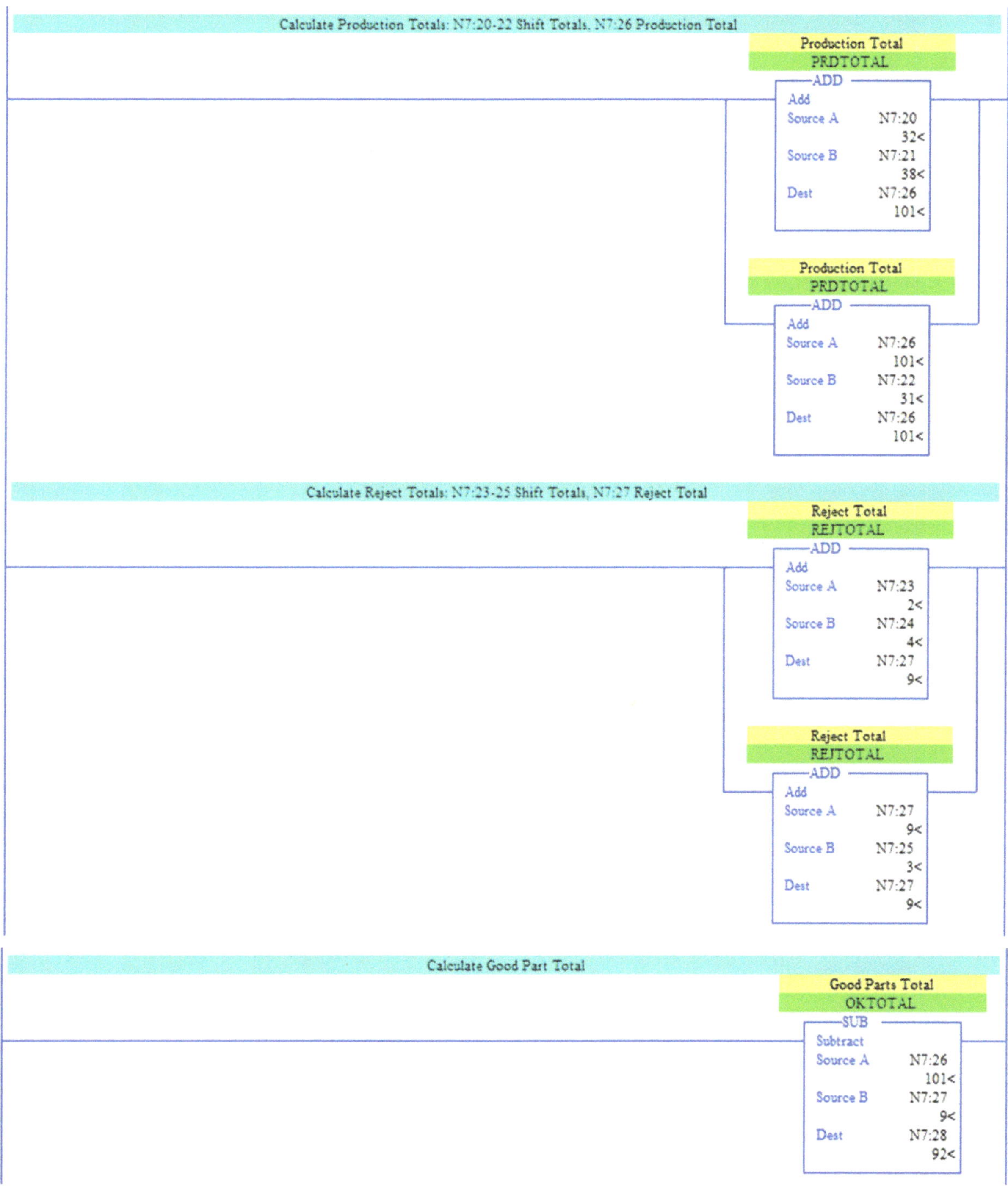

Exercise 10

1. A tank used for blending juice holds approximately 8,000 gallons of liquid. There is a
    pressure transducer that produces a 4-20mA signal, it is wired into channel 1 of an
    analog card.

    The tank is filled with 6,000 gallons of juice; the reading of the analog card is recorded as
    24,780. The tank is then drained completely and the value from the transducer is
    recorded as 96.

    Draw ladder logic that scales the raw reading from the transducer into gallons. There
    are 3.78541 liters in one gallon. Also calculate the number of liters.

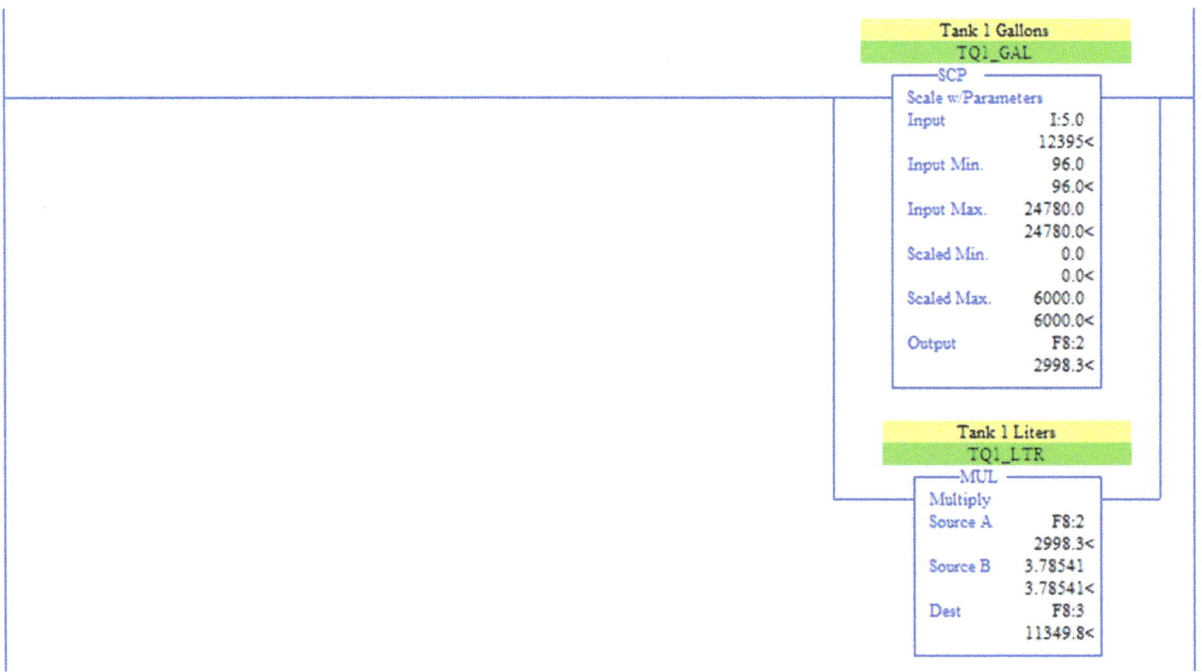

RSLogix500 - Program Structure and Advanced Topics

Program organization is done so that programmers can easily find areas of interest in the program. In RSLogix 500 platforms, the only routine that **must** be in the program is LAD 2, which is the main routine. It runs all the time. If all the code were placed in one routine however, it would make things very difficult to find and troubleshoot.

Likewise, RSLogix 500 programs start with one register of each data type; B, T, C, N, F and so on. As mentioned previously, up to 255 elements can be created in each register, but additional registers can also be created as needed also.

Routines (Ladders)

There are two common philosophies when creating routines for organization; that of creating routines by **function**, and that of creating routines by **area**. Combinations of these two methods can also be used.

Functional routines in a program include things like Input Conditioning, Outputs, Faults/Alarms, Auto Sequences or Control Routines, Production Data, and more. An important thing to remember when calling these routines is the order in which they are scanned: Input are updated at the beginning of the scan, the logic is evaluated, then the outputs are updated. For this reason, Inputs, Control Sequences and Outputs should be called in that order. Typically these are among the first routines called.

For larger programs a System routine is often created for mode control and things that are common to the entire machine or system. This keeps the Main routine (LAD 2) clear for subroutine calls. This will be discussed more completely in the next section on RSLogix Studio 5000.

It also only makes sense to create the routines in the order in which they are called. This makes it more intuitive for people evaluating the code. In RSLogix 500 the order of the LAD files cannot be changed, they will always appear in numerical order.

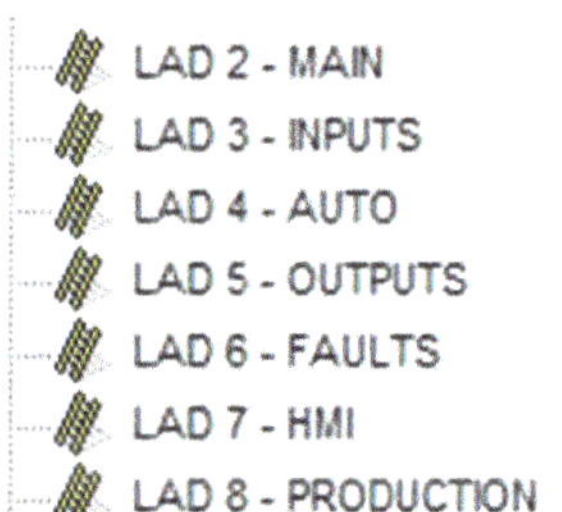

To the left is an example of a typical machine call structure. Ladders are called from the Main routine LAD 2 in the order in which they are listed. The system functions are located in the Main routine in this case. This is an example of organizing the routines by **function**.

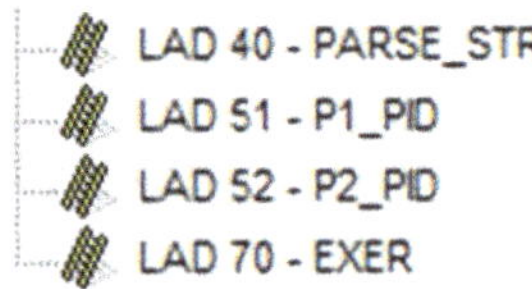

LAD 10 - TQ1 REG
LAD 11 - TQ1 CTRL
LAD 12 - TQ1 ALARM
LAD 13 - TQ2 REG
LAD 14 - TQ2 CTRL
LAD 15 - TQ2 ALARM
LAD 16 - TQ3 REG
LAD 17 - TQ3 CTRL
LAD 18 - TQ3 ALARM

In this case routines are organized by **area**, this is common in process control. It is still important that Inputs, Control and Outputs are called in that order, and they are, but they are all in the CTRL routines for each tank (TQ)

In this case each tank also has a Registration (REG) routine that handles permissives or interlocks for piping along with recipe and batch coordination. The ALARMS routines contain the alarms and faults for each tank.

Routines may also be called multiple times to accomplish a specific task.

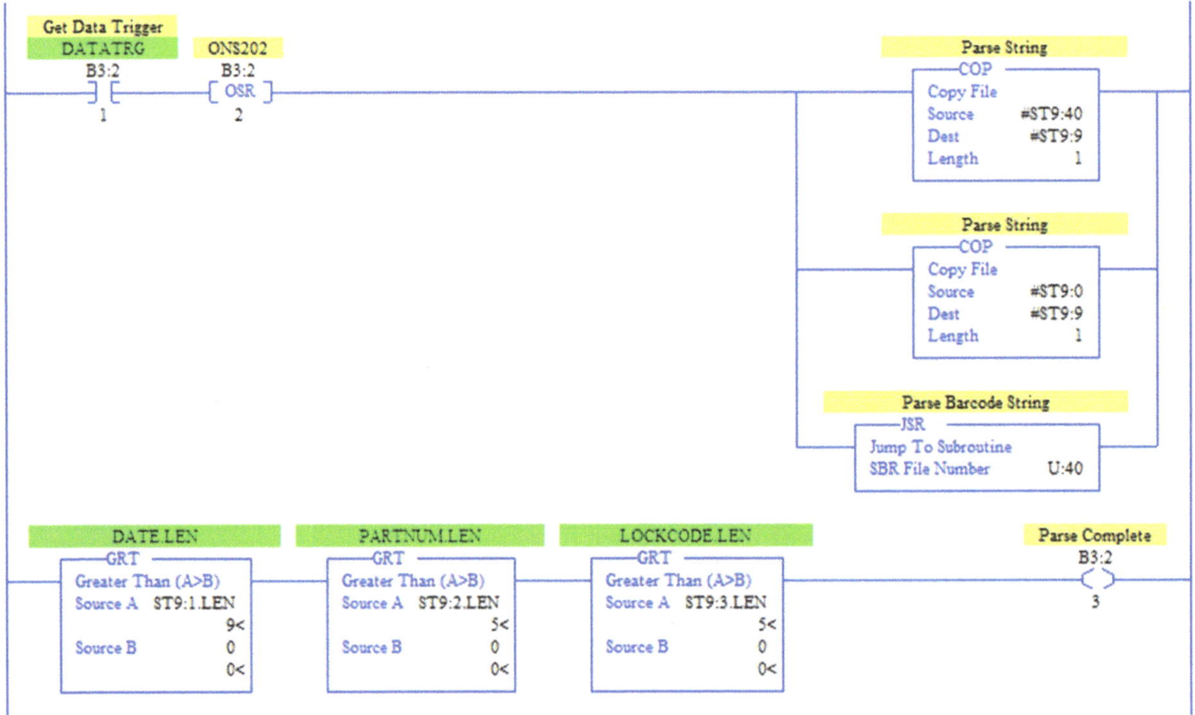

This code shows a string parsing routine that is called every time a barcode is received. ST9:9 is cleared with empty string ST9:40, received barcode ST9:0 is copied into ST9:9. Inside the subroutine the Date, Part Number and Lock Code are cleared, the string is "parsed" or disassembled into different string registers, and the data is used elsewhere.

LAD 40 - PARSE_STR
LAD 51 - P1_PID
LAD 52 - P2_PID
LAD 70 - EXER

The routines at left are either called outside of LAD 2 or are used to hold extraneous logic and not used at all in the program (LAD 70 - EXER). Routines 51 and 52 are called on a slower time schedule and are PID loops. They are placed at the end of the file list to illustrate that they are not in the continuous program scan.

Data Files

The first data files are already created by the software when the program is started. There is only one element in each file though, so it is important to plan the project structure before making new folders. For machinery and functional routines this is less important, but if the program is large or area-based with repetitive logic, it is useful to separate registers for easy duplication.

The list to the right shows a set of data files that have been created for the process tanks TQ1-3 shown previously. By creating separate bits, integers, REALs and Timers for each tank, routines can be duplicated and the addresses edited for each area. The Integer addresses for Tank 2 can be modified from N21 to N31 for every tag after copying the code from Tank 1. Notice also that space has been left for more registers if needed, it is not necessary to use all numbers.

Under **File** there is a selection for Data File that brings up the dialog below. This is a useful tool to view details of the files that have been created and modify them if necessary.

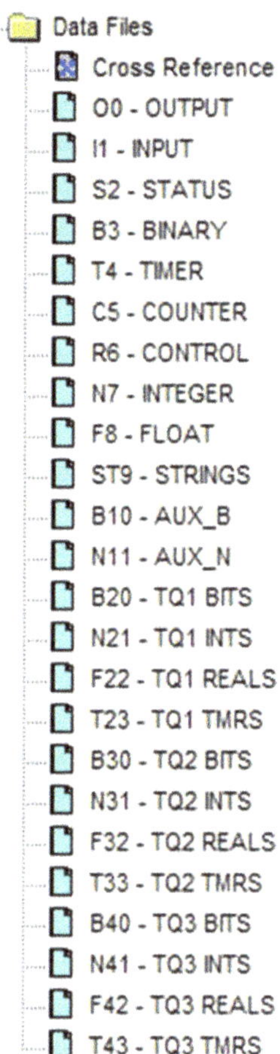

Note that as with the routines, file numbers do not need to be consecutive, but they will appear in numerical order. Keep this in mind when assigning them in the program.

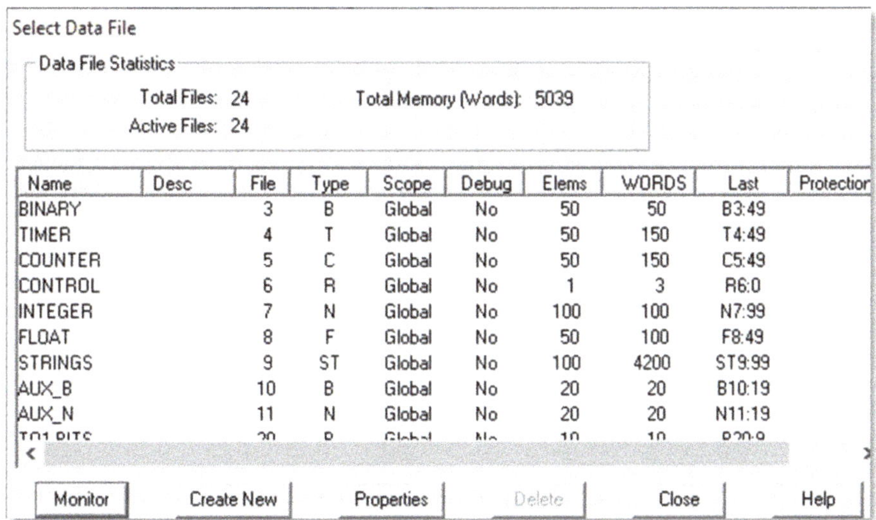

Name	Desc	File	Type	Scope	Debug	Elems	WORDS	Last	Protection
BINARY		3	B	Global	No	50	50	B3:49	
TIMER		4	T	Global	No	50	150	T4:49	
COUNTER		5	C	Global	No	50	150	C5:49	
CONTROL		6	R	Global	No	1	3	R6:0	
INTEGER		7	N	Global	No	100	100	N7:99	
FLOAT		8	F	Global	No	50	100	F8:49	
STRINGS		9	ST	Global	No	100	4200	ST9:99	
AUX_B		10	B	Global	No	20	20	B10:19	
AUX_N		11	N	Global	No	20	20	N11:19	
TQ1 BITS		20	B	Global	No	10	10	B20:9	

System Functions

System control is done in different ways depending on the process or machine being programmed. In **process control**, each unit (such as a pump or valve) may have its own mode of operation, such as Hand (Manual), Off, and Auto. This may be controlled by actual physical switches at the unit but may be overridden by the master system (such as SCADA) if local control might disrupt the process.

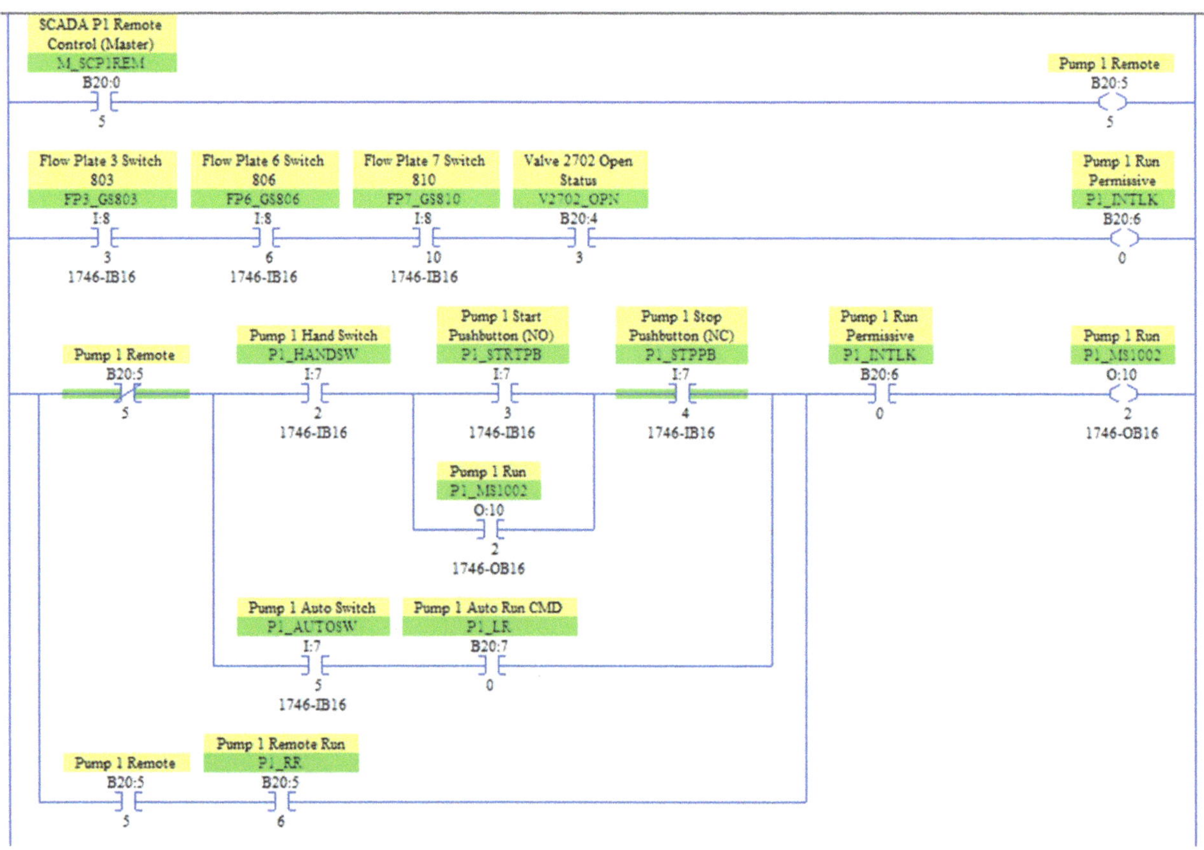

As shown above, there are two different ways that the system can operate. The Pump Remote Run signal may also incorporate manual control from an HMI or terminal.

Machine control usually treats the entire machine or zone as a single unit. The machine is placed in Auto or Manual mode, then the system is started from Auto by pressing a button for a short period of time, ensuring that personnel are aware that the machine is starting.

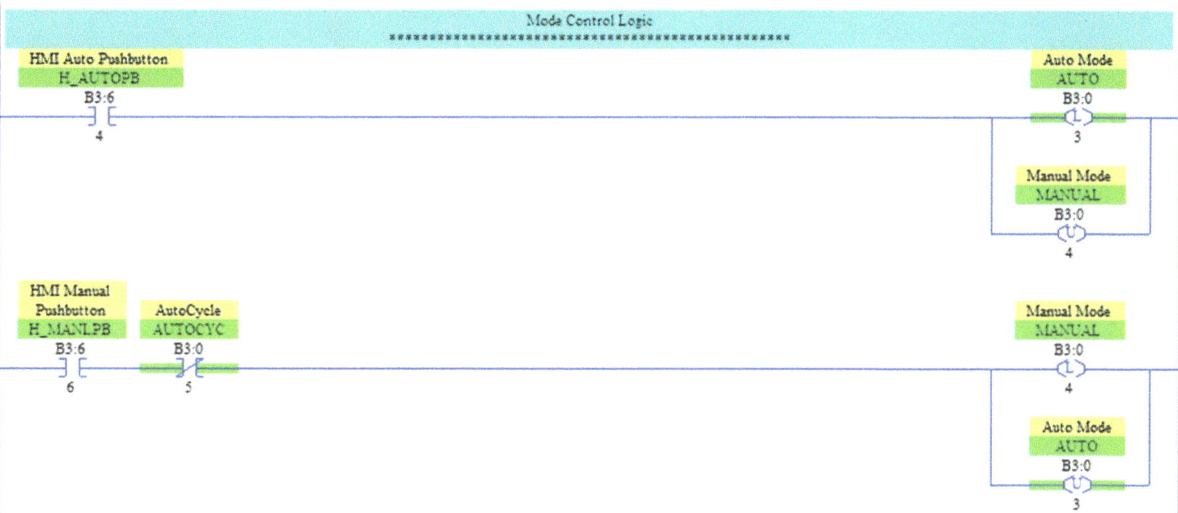

As shown above, Auto and Manual mode are usually pretty simple. The machine can't be placed in Manual until it is out of AutoCycle.

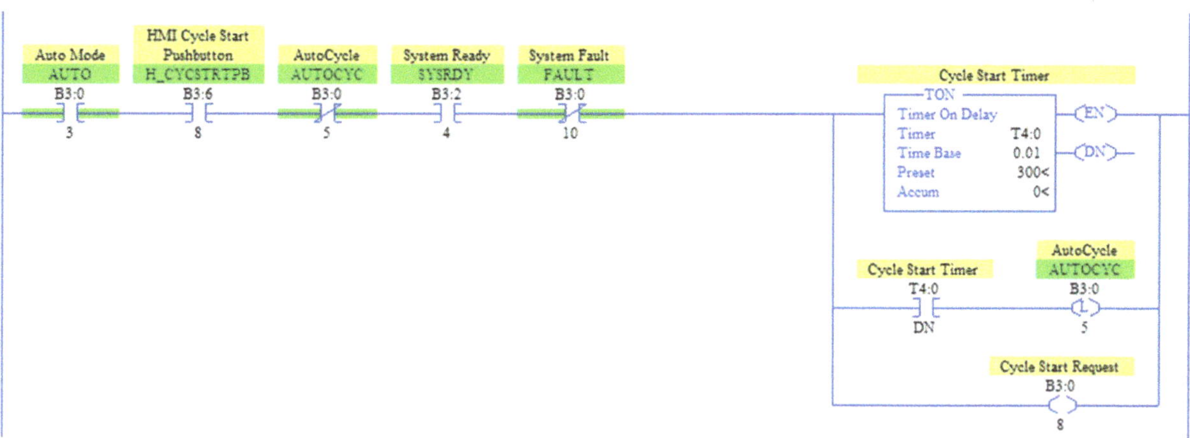

The logic above shows that the Cycle Start pushbutton needs to be held down for 3 seconds while in Auto Mode, with no faults and the system ready before AutoCycle is entered. The Cycle Start Request bit can be used to sound a horn or buzzer while the machine is started to alert personnel. This is also why the Normally Closed (NC) AutoCycle bit is placed in series, it prevents the horn from being sounded if the system is already running.

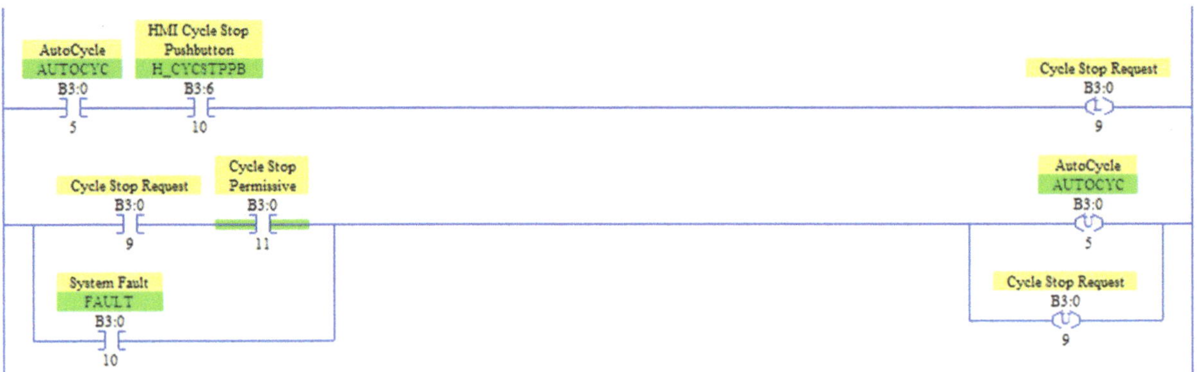

The above logic shows that the AutoCycle bit should not be stopped immediately but should wait until the permissive bit is true. This allows sequences and operations to come to a safe and repeatable stop before the cycle is ended. Faults or Immediate Stops will end the cycle immediately, but this can make it difficult to return a machine to its home position, or even damage product. If possible, all product operations should be complete and not remain in transit (such as grippers) before the cycle stops.

Other system functions are discussed in the CompactLogix and ControlLogix section.

Input Treatment

It is often useful to map inputs to status bits rather than using the input address directly. Two concepts that illustrate this are **Debouncing** of photoeyes, and **Back-Checking** complementary sensors.

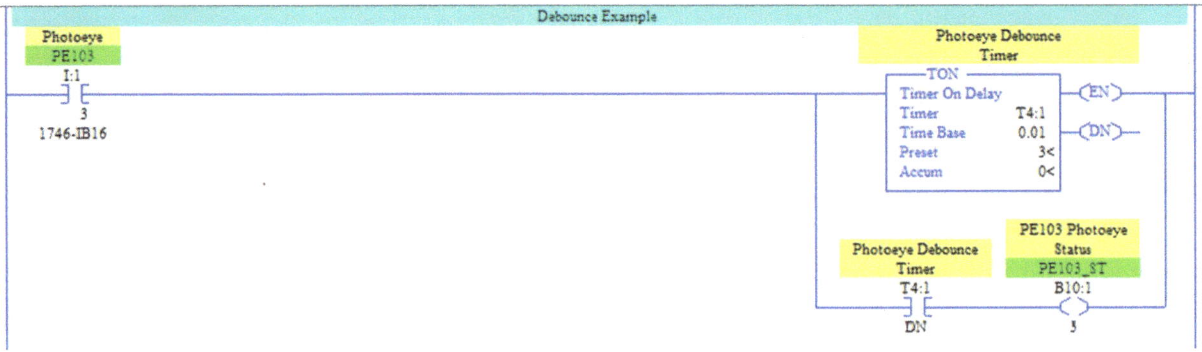

Applying timers to sensors is often call debouncing. The on-delay timer above shows that the sensor needs to be on for at least 30 milliseconds before the status bit is on, allowing the program to only evaluate the sensor when the detected part is stable and ignoring transient signals. This can be useful in ensuring that a part is at rest before acting on the signal.

Off-delay timers can also be used to "stretch" the signal, making the status bit stay on longer.

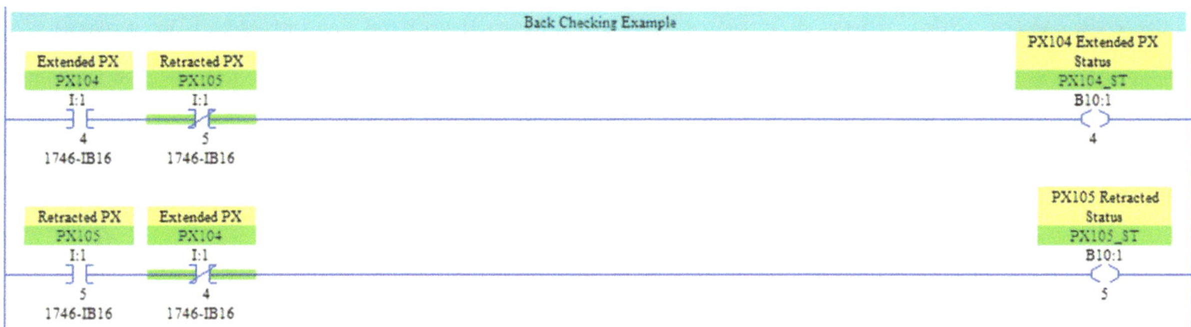

Sensors at opposite ends of an actuator can be "back checked" to ensure that their status bits never turn on at the same time. This is useful in Fault evaluation; if both sensors are on the status bit will stay off, triggering the timed fault bit shown later in this section.

Debouncing and backchecking can be used on the same status bit as required. In the ControlLogix section an "AOI" will be shown that illustrates this.

An added benefit of using status bits is that all inputs are listed in the input routine for troubleshooting. For this reason, it might be useful to map <u>all</u> physical inputs to status bits.

Outputs

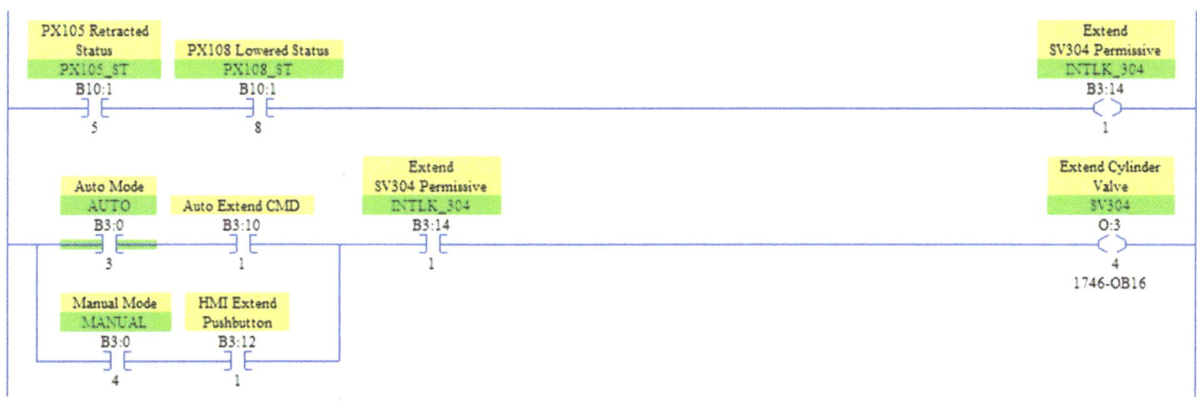

Output structures are typically all located in the same routine for machine control in RSLogix 500, at least for smaller projects. They are usually listed in numerical order of the addresses so that they can be easily located for troubleshooting.

Permissives or interlocks are usually placed above the output coil for the same reason. If an output does not activate, it is often due to a permissive not being satisfied; this makes it easy to find.

The main purpose for permissives is to prevent damage to an actuator or other part of the machine. For this reason, other run conditions like faults or machine-wide conditions are usually not placed in the permissive rung. The interlock above shows that the cylinder is not allowed to extend unless the two listed proxes (proximity switches) are activated. This also works with the back-checking sensor status described previously; if a sensor fails in the on state it will fault the machine.

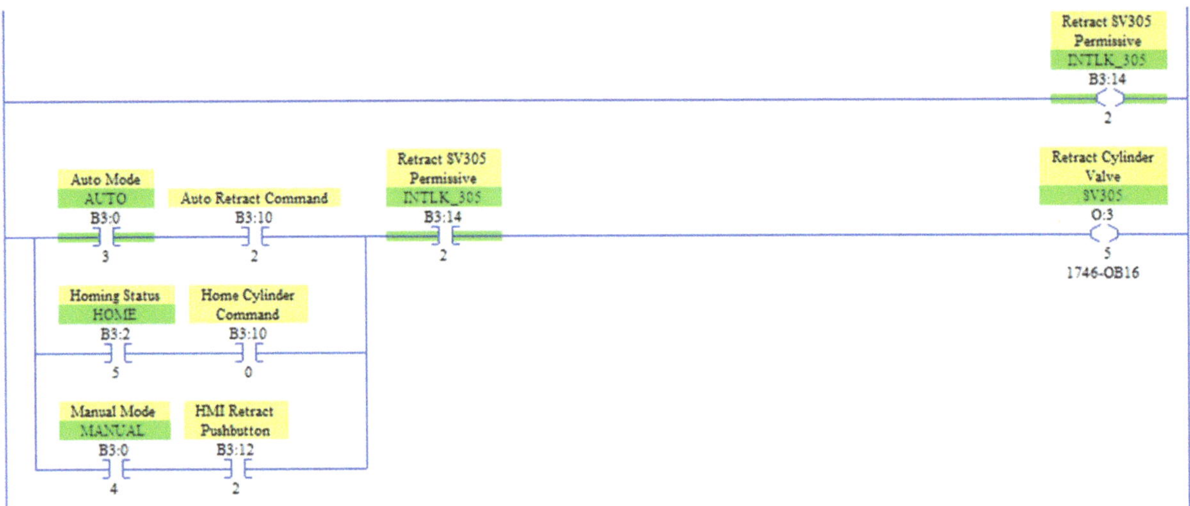

Interlocks can be written with no contacts in front of them making them "always on", this is not true on some other platforms.

There can be multiple conditions that activate an output coil as shown above. Auto and Manual modes are very common with a pushbutton activating the output in Manual. Homing of a system is another condition, which may occur in Auto or Manual, or in a mode of its own. The permissive applies in all cases.

**It is very important that coils are never in more than on location in the program!**

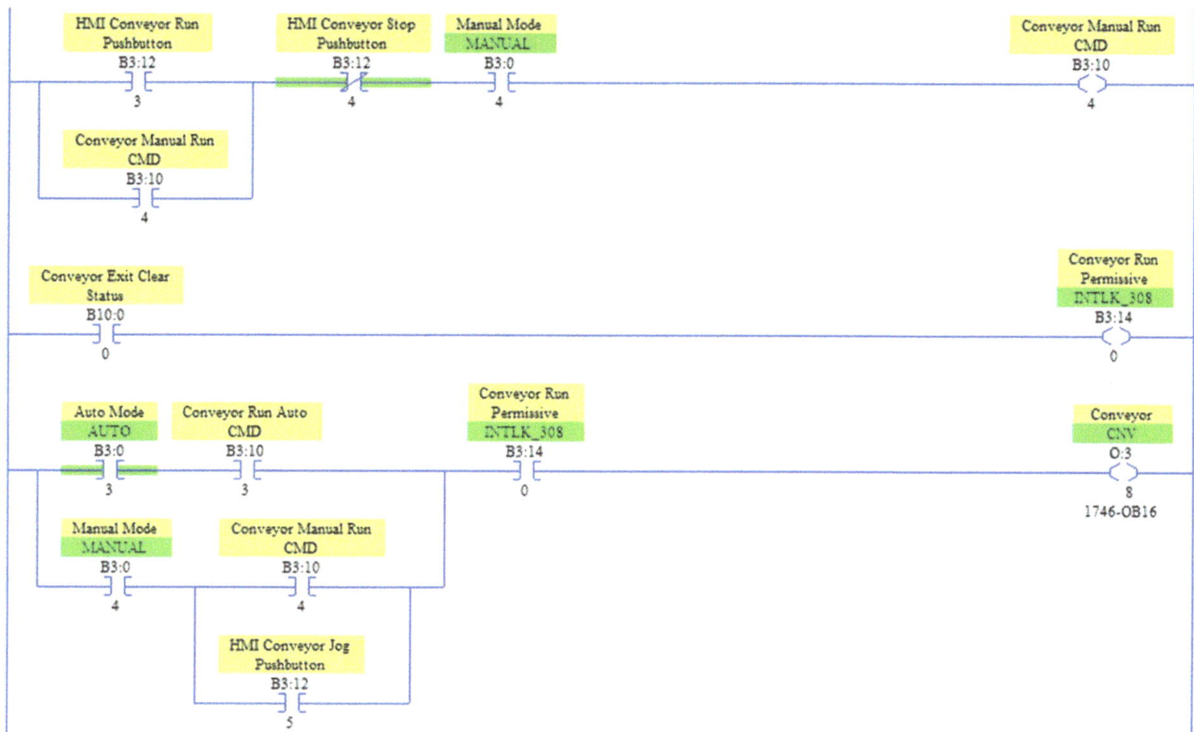

Air cylinders are straightforward to control in Manual Mode: The HMI button directly controls the valve. Usually, the mechanics of the valve itself will control whether the cylinder reaches its full extent based on the action of the internal spool. Conveyors are less so; they may need to stay on or be able to be jogged. The logic above shows a method of doing this. Notice that the Manual Run CMD bit drops out when the system is placed in Auto. This is also why the Jog button is not placed in the logic for the CMD.

It is also important to use the Auto Mode bit in the output structure rather than the AutoCycle bit. This allows outputs to remain on when the system faults. If it is necessary to turn an output off when it faults, place the fault in series with the Auto CMD in the control routine.

Faults and Alarms

Problems occurring in machine or system control are usually called faults, which in turn create alarms that alert an operator or technician to the problem. Faults then serve two important purposes: to stop the machine from operating, preventing damage to itself, product or personnel, and to alert someone to come and correct the problem.

Nearly any abnormal condition can be detected by programming, but defining the problem can be more difficult. Following are some standard types of fault detection methods.

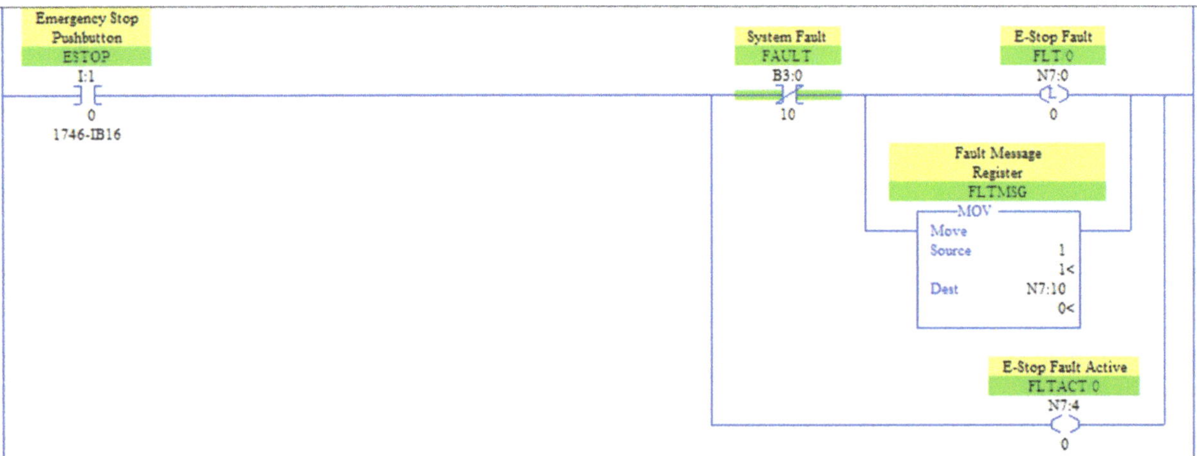

Above is one of the simplest types of fault. If the normally open contact on an E-Stop pushbutton is activated, it means that someone has pressed the E-Stop, removing power from actuators by disengaging the supply to an output card. This power removal is done by dropping power to a relay called the MCR, or Master Control Relay through two sets of NC contacts on the same pushbutton. The fault then does not stop the output, the removal of power does.

Note that there are two registers affected by the button; the FLT register and the FLTACT register. The FLT register is latched in, signifying that the fault has occurred and has not been cleared (memory), while the FLTACT (Fault Active) register shows whether the cause of the fault is still present. These are generally groups of bits from the same integer, making them easier to track and clear.

The FLTMSG (Fault Message) register is used to hold the number of the message that will be displayed to the operator. Moving this number into a register in the HMI will display that message number from a list.

Note also that the FAULT (System Fault) message has been placed in series with the latch and the message movement. This prevents other faults resulting from pressing the E-Stop from activating also.

It is also possible to display multiple fault messages by scrolling them on the HMI display, but often operators want to see the initial cause of the problem.

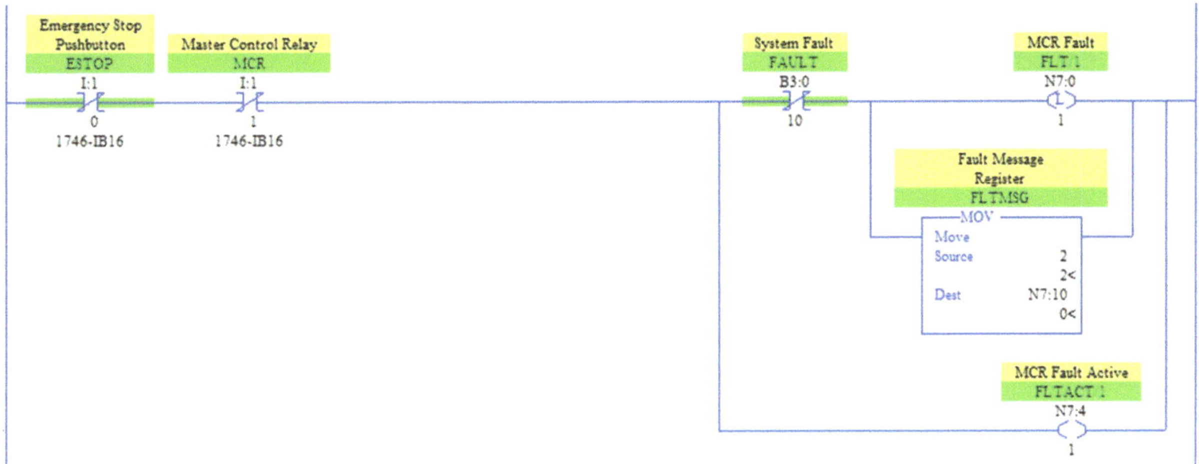

The fault above is cause by the E-Stop button not being pressed, but the MCR not being active. The MCR is usually reset by pressing a power/reset button, but if there is a problem with the circuit the relay may not reset. This is often caused by a problem in one of the MCR channels.

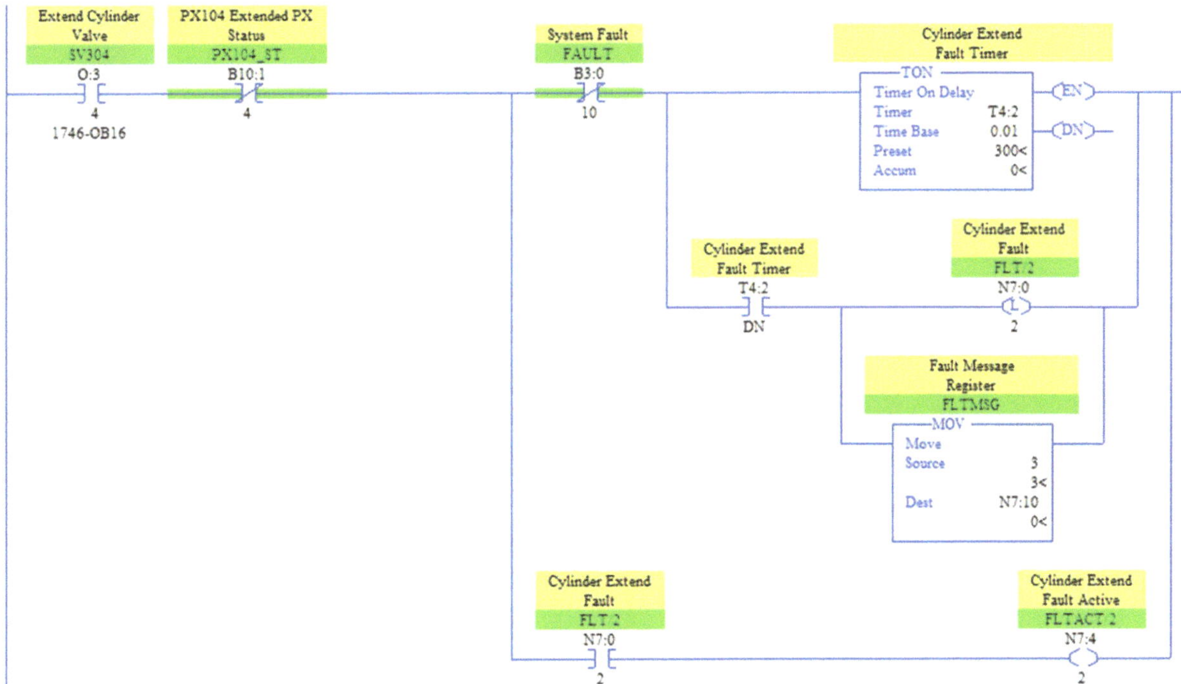

Timed faults are common with actuators; if an output is activated but there is no response within a period of time, a fault is activated. This is a case when a specific fault might be placed in the CMD logic to prevent the actuator from continuing to try and reach the sensor. If this is done, the fault active bit will also go off and the timer would start timing from zero,

An "Air Dump" is also sometimes used to remove air from cylinders.

Note that the Status (ST) bit resulting from the back checking of the sensor is used here, rather than the direct input address. This allows the fault to be detected if a sensor fails in either state.

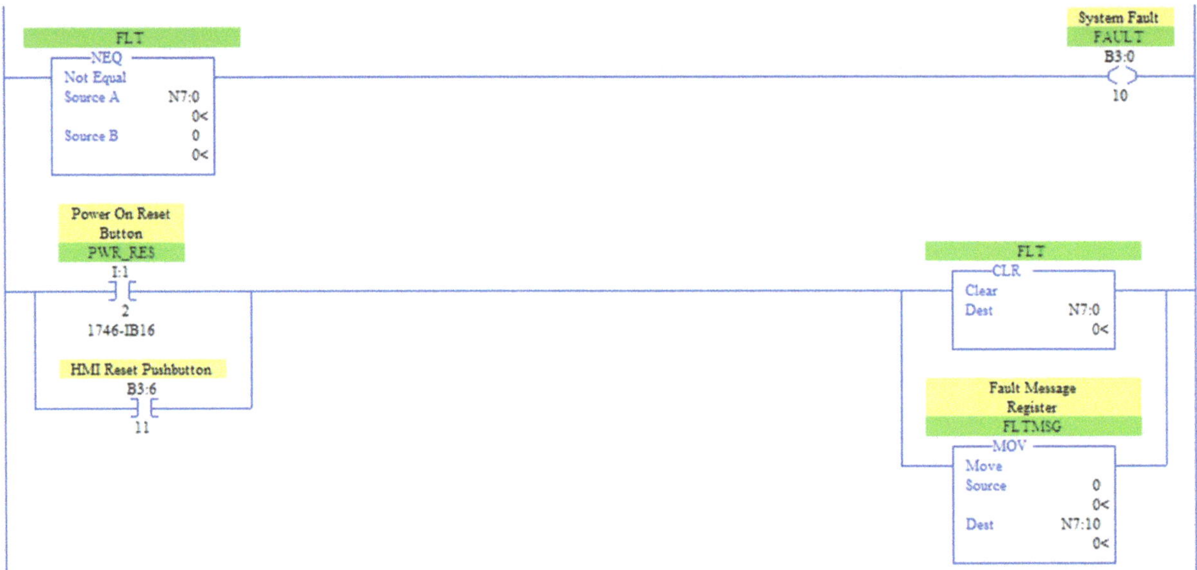

Because the fault bits (FLT) were all in the same integer, it is easy to use an NEQ instruction to determine if a fault is present. The CLR instruction resets the FLT register to zero, unlatching all of the bits in the integer, and the MOV of 0 to the FLTMSG register places a "No Faults Present" message in the HMI. If necessary, detecting whether the fault is still active can be used to disable the reset, perhaps with a Fault Reset Permissive bit.

Control and Sequences

Simple control of actuators and devices can be done as shown in the Outputs section. If it is desired to have a conveyor run any time the system is in Autocycle, the Auto CMD output bit can be made active whenever the system is running.

If it is desired to have the device run only under certain conditions or in a certain order, it is necessary to define the conditions accurately. If the actuators need to work together, one of the easiest methods is to use an **Auto Sequence** or a **State Machine**.

Two modern methods of writing an Auto Sequence are described in the ControlLogix section of this book. These methods, Bit Sequences and Word Sequences, can be applied using the N7 registers in RSLogix 500.

Allen-Bradley CompactLogix and ControlLogix Platforms

Rockwell Software – RSLogix 5000 and Studio 5000

Until version 20, Allen-Bradley's CompactLogix and ControlLogix programming software was named RSLogix 5000. As of version 20.011, (about 2014), the software was renamed Studio 5000 Logix Designer; many newer products are not supported by the older RSLogix 5000.

Name	Catalog #	Controllers	Description
Service Edition	9324-RLD000ENE	CompactLogix5370 ControlLogix5500 SoftLogix5800 Compact GuardLogix5300 GuardLogix5500	Upload/ Download and View Only
Mini Edition	9324-RLD200ENE	CompactLogix5370 Compact GuardLogix5300	Ladder Diagram (LD) is fully supported. Function Block Diagram (FBD), Sequential Function Chart (SFC), and Structured Text (ST) is not supported.
Lite Edition	9324-RLD250ENE	CompactLogix5370 Compact GuardLogix5300	Ladder Diagram (LD), Function Block Diagram (FBD), Sequential Function Chart (SFC), and Structured Text (ST) is supported.
Standard Edition	9324-RLD300ENE	CompactLogix5370 ControlLogix5500 SoftLogix5800 Compact GuardLogix5300 GuardLogix5500	Ladder Diagram (LD) is fully supported. Function Block Diagram (FBD), Sequential Function Chart (SFC), and Structured Text (ST) is not supported.
Full Edition	9324-RLD600ENE	CompactLogix5370 ControlLogix5500 SoftLogix5800 Compact GuardLogix5300 GuardLogix5500	Ladder Diagram (LD), Function Block Diagram (FBD), Sequential Function Chart (SFC), and Structured Text (ST) is supported.
Professional Edition	9324-RLD700NXENE	CompactLogix5370 ControlLogix5500 SoftLogix5800 GuardLogix5500	Ladder Diagram (LD), Function Block Diagram (FBD), Sequential Function Chart (SFC), and Structured Text (ST) is supported and includes RSNetWorx for ControlNet, DeviceNet, EtherNet/IP (9357-CNETL3, 9357-DNETL3, 9357-ENETL3 individually or 9357-ANETL3 combined), and RSLogix Emulate 5000.

Latest version as of April 2023 is Version 35.

1769 CompactLogix 5370 Controllers

Software:

RSLogix 5000 or Studio 5000, Standard, Mini or Lite Editions

Models:

5370 L3: Max I/O points 960, 3MB User Memory, Max Motion Position Axes 16. Uses 1769 Compact I/O.

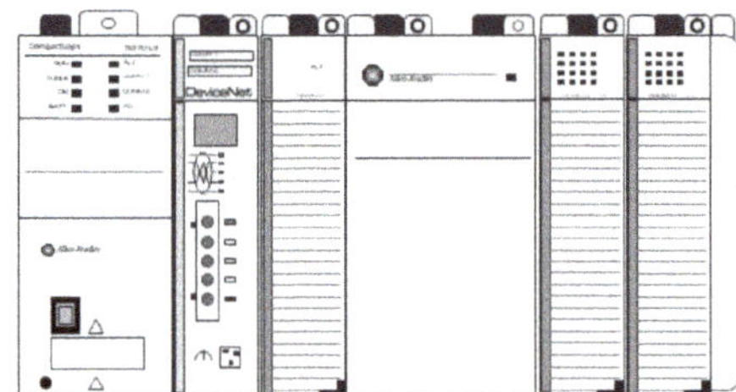

5370 L2: Max I/O points 160, 1MB User Memory, Max Motion Position Axes 4. Uses 1769 Compact I/O.

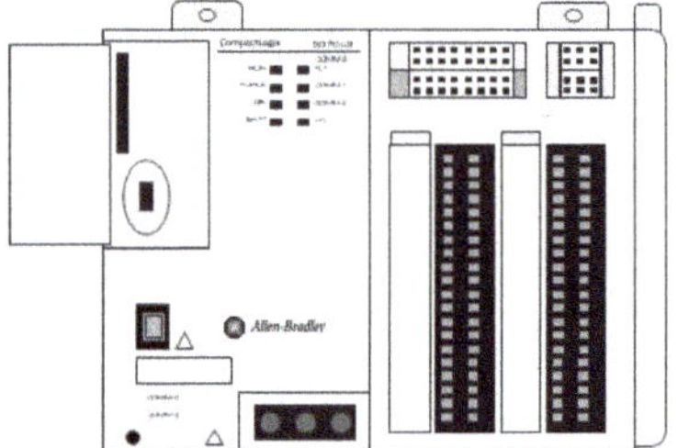

5370 L1: Max I/O points 96, 1MB User Memory, Max Motion Position Axes 2. Uses 1734 Point I/O. Has embedded power supply and I/O.

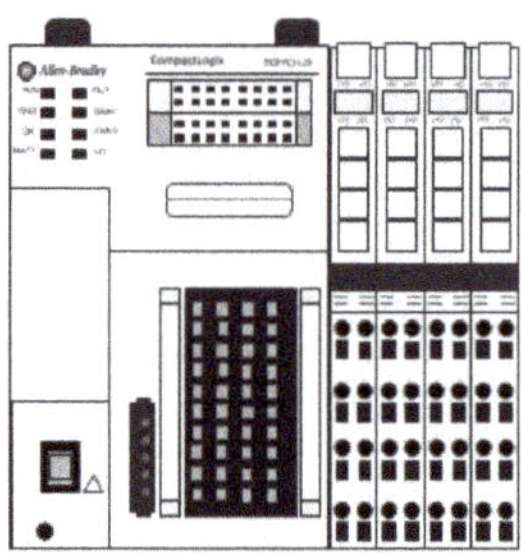

Compact GuardLogix 5370: Up to 960 I/O points, Standard 1, 2 or 3MB, Safety 0.5, 1 or 1.5 MB. 4, 8 or 16 Motion Position Axes. Uses 1769 Compact I/O. Up to 30 expansion Modules.

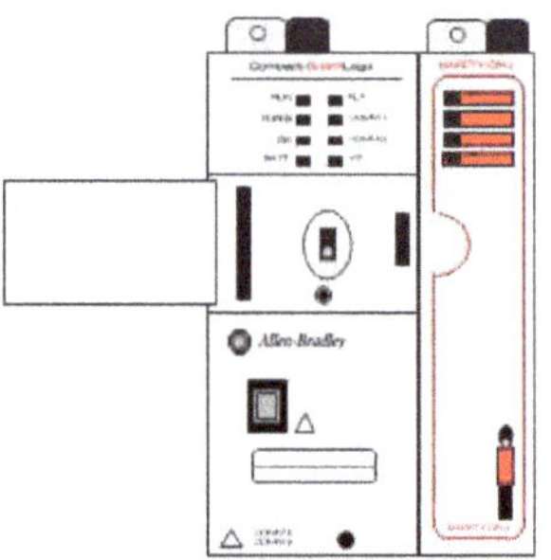

5069 CompactLogix 5380 Controllers

Software:

Studio 5000 Logix Designer, Standard, Mini or Lite Editions

Models:

5069-L3XX These controllers enable high-speed I/O, motion control, dual configurable Ethernet ports that allow dual IP addresses, enhanced diagnostics and troubleshooting. Integrated motion on Ethernet/IP is available up to 32 axes depending on the model.

Catalog Number	Application Memory	I/O Expansion	Ethernet Nodes	Motion Axes
5069-L306ER	0.6 MB	8	16	0
5069-L310ER	1 MB	8	24	0
5069-L320ER	2 MB	16	40	0
5069-L330ER	3 MB	31	50	0
5069-L340ER	4 MB	31	55	0
5069-L310ER-NSE	1MB	8	24	0
5069-L306ERM	0.6 MB	8	16	2
5069-L310ERM	1 MB	8	24	4
5069-L320ERM	2 MB	16	40	8
5069-L330ERM	3 MB	31	50	16
5069-L340ERM	4 MB	31	55	20
5069-L350ERM	5MB	31	60	24
5069-L380ERM	8MB	31	70	28
5069-L3100ERM	10MB	31	80	32

5069 CompactLogix 5480 Controllers

Software:

Studio 5000 Logix Designer, Version 35. Available in 2023.

Models:

5069-L4XX Logix based real-time controller that runs on Windows 10 IOT Enterprise. Enables high speed I/O, motion control, Device-level Ring/Linear topologies. Includes three GbE Ethernet/IP ports; two are configurable and one is a dedicated commercial OS network interface. Includes integrated DisplayPort for high-definition monitor connectivity, enhanced security features and two USB 3.0 ports for OS peripheral and expanded data storage capability. Supports up to 31 local Bulletin 5069 Compact I/O modules.

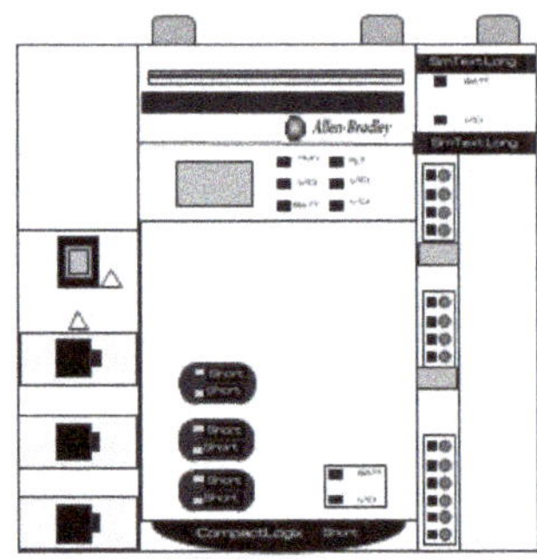

1768 CompactLogix L4x and L4xS Controllers

Software:

Studio 5000 Logix Designer, Standard, Mini or Lite Editions

Models:

1768-L4x These controllers combine both a 1768 backplane and a 1769 backplane. The 1768 backplane supports the 1768 controller, the 1768 power supply and a maximum of four 1768 modules. The 1769 backplane supports up to 16 1769 modules. All controllers have Ethernet/IP and RS-232 Port.

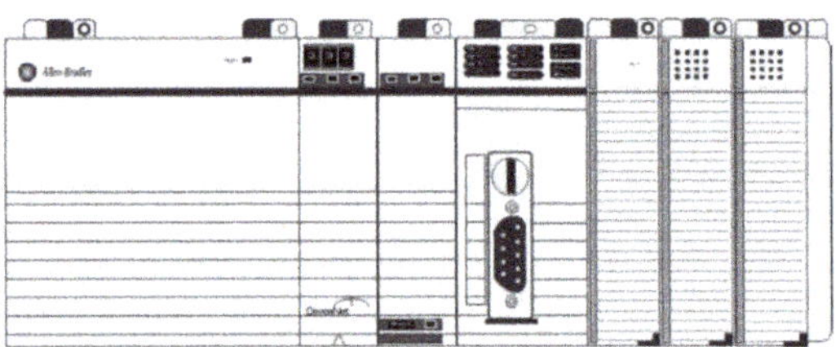

Catalog Number	User Memory
1768-L43	2 MB
1768-L43S	2MB + 0.5 MB Safety
1768-L45	3MB
1768-L45S	3MB + 1MB Safety

5069 CompactLogix 5480 Controllers

ControlLogix 5570 and 5580 Controllers

Software:

Studio 5000 Logix Designer, Ladder Logic, Function Block Diagram, Structured Text and Sequential Function Chart.

Models:

1756-L7X, 1756-L8X Rack-based system for larger projects. Rack sizes are 4, 7, 10, 13 and 17 slots.

The 1756-L6x series is not supported after the release of Studio 5000 Logix Designer; version 20 of RSLogix5000 is the last firmware revision.

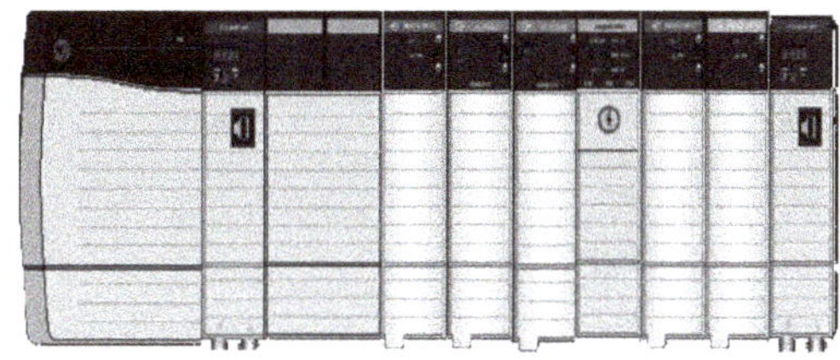

Catalog Number	User Memory	Catalog Number	User Memory
1756-L71	2 MB	1756-L81	3 MB
1756-L72	4 MB	1756-L82	5 MB
1756-L73	8 MB	1756-L83	10 MB
1756-L74	16 MB	1756-L84	20 MB
1756-L75	32 MB	1756-L85	40 MB

The L7X series has a USB Port, while the L8X series has an additional 1GB Ethernet port. An alphanumeric display shows processor status and name. Supports Integrated Motion on Ethernet/IP, full controller redundancy and Removal Insertion Under Power.

GuardLogix Safety Controllers (1756-LXX-S) provide safety and integrated motion in the same chassis, while Extreme Environment Controllers (1756-LXX-XT) allow operation from -25 to 70 C, or -13 to 158 F.

CompactLogix and ControlLogix Instructions

Basic Instructions

Mnemonic	Name	Purpose
XIC	Examine If Closed, Examine On	Examines a bit for an ON condition
XIO	Examine If Open, Examine Off	Examines a bit for an OFF condition
OTE	Output Energize	Turns a bit ON or OFF
OTL	Output Latch	Sets a bit ON when executed, the bit retains its state until unlatched or the register is cleared
OTU	Output Unlatch	Resets a bit OFF when executed
OSR	One-Shot Rising	Triggers a one time event on leading edge of signal, ON for one scan
OSF	One-Shot Falling	Triggers a one time event on falling edge of signal, ON for one scan
ONS	One-Shot (Rising)	Same as OSR
TON	Timer On-Delay	Counts timebase intervals when the energizing instruction is TRUE
TOF	Timer Off-Delay	Counts timebase intervals when the energizing instruction is FALSE
RTO	Retentive Timer	Counts timebase intervals when the energizing instruction is TRUE, retains accumulated value when energizing instruction is FALSE
CTU	Count Up	Increments the accumulated value at each false to true transition and retains the accumulated value when the instruction goes false or when power cycle occurs
CTD	Count Down	Decrements the accumulated value at each false to true transition and retains the accumulated value when the instruction goes false or when power cycle occurs
HSC	High-speed Counter	Counts high-speed pulses from a fixed controller high-speed input
RES	Reset	Resets the accumulator value and status bits of a timer or counter. *Do not use with TOF Timers!

Comparison Instructions

Mnemonic	Name	Purpose
EQU	Equal	Test whether two values are equal
NEQ	Not Equal	Test whether two values are not equal
LES	Less Than	Test whether one value is less than another value
LEQ	Less Than or Equal	Test whether one value is less than or equal to another value
GRT	Greater Than	Test whether one value is greater than another value
GEQ	Greater Than or Equal	Test whether one value is greater than or equal to another value
MEQ	Masked Equal	Test portions of two values to determine whether they are equal through a mask
LIM	Limit Test	Test whether one value is within the range of two other values

Math Instructions

Mnemonic	Name	Purpose
ADD	Add	Adds source A to source B and stores the result in the destination
SUB	Subtract	Subtracts source B from source A and places the result in the destination
MUL	Multiply	Multiplies source A by source B and places the result in the destination
DIV	Divide	Divides source A by source B and places the result in the destination
DDV	Double Divide	Divides the contents of the math register by the source and stores the result in the destination and the math register
CLR	Clear	Sets all bits of a word to zero
SQR	Square Root	Calculates the square root of the source and places the result in the destination
SCP	Scale with Parameters	Produces a scaled output value that has a linear relationship between the input and scaled values
SCL	Scale Data	Multiplies the source by a specific rate, adds to an offset value, and stores the result in the destination

RMP	Ramp	Provides the ability to create linear acceleration, deceleration and "S" curve ramp output data waveforms
ABS	Absolute	Calculates the absolute (positive) value of the source and places the result in the destination
CPT	Compute	Evaluates an expression and places the result in the destination
SWP	Swap	Swaps the low and high bytes of a specified number of words in a bit, integer, ASCII or string file
COS	Cosine	Takes the cosine of a number and stores the result in the destination
SIN	Sine	Takes the sine of a number and stores the result in the destination
TAN	Tangent	Takes the tangent of a number and stores the result in the destination
ASN	Arc Sine	Takes the arc sine of a number and stores the result (in radians) in the destination
ACS	Arc Cosine	Takes the arc cosine of a number and stores the result (in radians) in the destination
ATN	Arc Tangent	Takes the arc tangent of a number and stores the result (in radians) in the destination
LN	Natural Log	Takes the natural log of the value in the source and stores the result in the destination
LOG	Log to the base 10	Takes the log base 10 of the value in the source and stores the result in the destination

Data Handling Instructions

Mnemonic	Name	Purpose
TOD	Convert to BCD	Converts the integer source value to BCD format and stores it in the destination
FRD	Convert from BCD	Converts the BCD source value to an integer and stores it in the destination
DEG	Convert from Radians to Degrees	Converts radians (source) to degrees and stores the result in the destination
RAD	Convert from Degrees to Radians	Converts degrees (source) to radians and stores the result in the destination
DCD	Decode 4 to 1 of 16	Decodes a 4 bit value (0 to 15), turning on the corresponding bit in the 16 bit destination

ENC	Encode 1 of 16 to 4	Encodes a 16 bit source to a 4 bit value. Searches the source from the lowest to the highest bit and looks at the first set bit. The corresponding bit position is written to the destination as an integer.
COP	Copy File	Copies data from the source file to the destination file
FLL	Fill File	Loads a source value into each position in the destination file
MOV	Move	Copies the source value to the destination
MVM	Masked Move	Copies the source value to part of the destination
AND	And	Performs a bitwise AND operation
OR	Or	Performs a bitwise OR operation

XOR	Exclusive Or	Performs a bitwise inclusive OR operation
NOT	Not	Performs a NOT (invert) operation
NEG	Negate	Changes the sign of the source and stores it in the destination
FFL	FIFO Load	Loads a word into a FIFO (First In - First Out) stack on each false to true transition. The first word loaded is the first to be unloaded.
FFU	FIFO Unload	Unloads a word from a FIFO (First In - First Out) stack on each false to true transition. The first word loaded is the first to be unloaded.
LFL	LIFO Load	Loads a word into a LIFO (Last In - First Out) stack on each false to true transition. The last word loaded is the first to be unloaded.
LFU	LIFO Unload	Unloads a word from a LIFO (Last In - First Out) stack on each false to true transition. The last word loaded is the first to be unloaded.

Program Flow Instructions

Mnemonic	Name	Purpose
JMP, LBL	Jump to Label and Label	Jump forward or backward to a specified "Label" instruction
JSR	Jump to Subroutine	Jump to a designated subroutine or ladder
SBR	Subroutine label	Designates the start of a subroutine or ladder
RET	Return from subroutine	Returns from a subroutine to the point from which it was called
MCR	Master Control Reset	Turn off all non-retentive outputs in a section of ladder
TND	Temporary End	Mark a temporary end that halts program execution
SUS	Suspend	Identifies specific conditions for program debugging and system troubleshooting
IIM	Immediate Input with Mask	Program an immediate input update using a mask
IOM	Immediate Output with Mask	Program an immediate output update using a mask
REF	Refresh	Interrupt the program scan to update the I/O and service communications

Application Specific Instructions

Mnemonic	Name	Purpose
BSL, BSR	Bit Shift Left or Right	Loads a bit of data into a bit array, shifts the pattern through the array and unloads the last bit of data in the array. BSL shifts data to the left, while BSR shifts it to the right
SQO, SQC	Sequencer Output and Sequencer Compare	Controls sequential machine operations by transferring 16 bit data through a mask to image addresses
SQL	Sequencer Load	Captures referenced conditions by manually stepping the machine through its operating sequences
TDF	Compute Time Difference	Calculates the number of 10 microsecond ticks between any two captured time stamps
FBC	File Bit Compare	Compares bits between two different files

DDT	Diagnostic Detect	Used to monitor machine or process operations to detect malfunctions
RPC	Read Program Checksum	Copies the program checksum from processor memory or from the memory module into the data table

ASCII Instructions

Mnemonic	Name	Purpose
ABL	Test ASCII Buffer for Line	Determine the number of characters in the buffer up to and including the user configured end of line characters
ACB	Number of ASCII Characters in Buffer	Determine the total number of characters in the buffer
ACI	String to Integer	Convert a string to an integer value
ACL	ASCII Clear Buffer	Clear the receive and /or transmit buffers
ACN	String Concatenate	Combine two strings into one
AEX	String Extract	Extract a portion of a sring to create a new string
AHL	ASCII Handshake Lines	Set or reset modem handshake lines
AIC	Integer to String	Convert an integer value into a string
ARD	ASCII Read Characters	Read characters from the input buffer and place them into a string
ARL	ASCII Read Line	Read one line of characters from the input buffer and place them into a string
ASC	String Search	Search a string
ASR	ASCII String Compare	Compare two strings
AWA	ASCII Write with Append	Write a string with user configured characters appended
AWT	ASCII Write	Write a string

Block Transfer and PID Instructions

Mnemonic	Name	Purpose
BTR	Block Transfer Read	Receive data from a remote device
BTW	Block Transfer Write	Send data to a remote device
PID	Proportional Integral Derivative	Controls physical properties such as temperature, pressure, liquid level or flow rate using closed process loops

Interrupt Routine Instructions

Mnemonic	Name	Purpose
	User Fault Routine	Provides the option of preventing a processor shutdown
STI	Selectable Timed Interrupt	Allows you to interrupt the scan of the main program file automatically, on a periodic basis, to scan a specified subroutine file
STD	Selectable Timed Disable	Disables STIs from occurring
STE	Selectable Timed Enable	Enables STIs to occur
STS	Selectable Timed Start	Sets or changes the file number or setpoint frequency of the STI routine
DII	Discrete Input Interrupt	Allows the processor to execute a subroutine when the input pattern of a discrete input card matches a compare value that you programmed.
ISR	I/O Interrupt	Allows a specialty I/O module to interrupt the normal processor operating cycle in order to scan a specific subroutine file
IID	I/O Interrupt Disable	Disables I/O Interrupts from occurring
IIE	I/O Interrupt Enable	Enables I/O Interrupts to occur
RPI	Reset Pending Interrupt	Aborts a pending I/O Interrupt
INT	Interrupt Subroutine	Optional instruction to identify interrupt subroutines

Communication Instructions

Mnemonic	Name	Purpose
SVC	Service Communications	Interrupts the program scan to execute the service communication portion of the operating cycle
MSG	Message Read/Write	Transmits data from one node to another on the network
CEM	ControlNet Explicit Message	Transmits CIP generic commands to other ControlNet nodes via the 1747-SCNR
DEM	DeviceNet Explicit Message	Transmits CIP generic commands to other DeviceNet nodes via the 1747-SDN
EEM	EtherNet/IP Explicit Message	Transmits CIP generic commands to other Ethernet/IP nodes via channel 1

For detailed information see Allen-Bradley Instruction Set Reference Manual 1747-rm001_-en-p

Starting and Editing a Project with RSLogix 5000

  Whether using RSLogix 5000 (version 19 and earlier) or Studio 5000 Logix Designer (version 20.011 and later), the software is installed in the Rockwell Software folder. There may also be an icon on the desktop as a shortcut.

Double-clicking on one of the icons will open the software environment. It is important to know ahead of time the firmware of your PLC controller, since these two versions can only be used to create projects within their scope. If an existing project is opened, the correct version will automatically be selected.

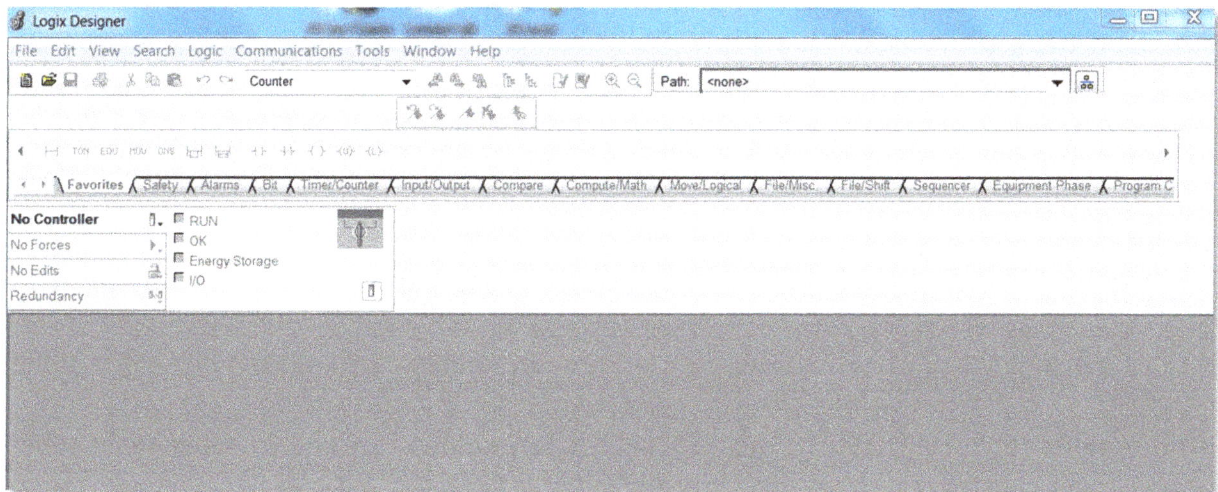

The programming environment looks like the figure shown above. This picture shows that the software selected was Studio 5000 as indicated by the "Logix Designer" title, while the image on the following page is titled "RSLogix 5000".

Creating a Project

To start a new project, a controller needs to be selected and named. This is done from the "File" tab. After selecting "New", the following screen will appear:

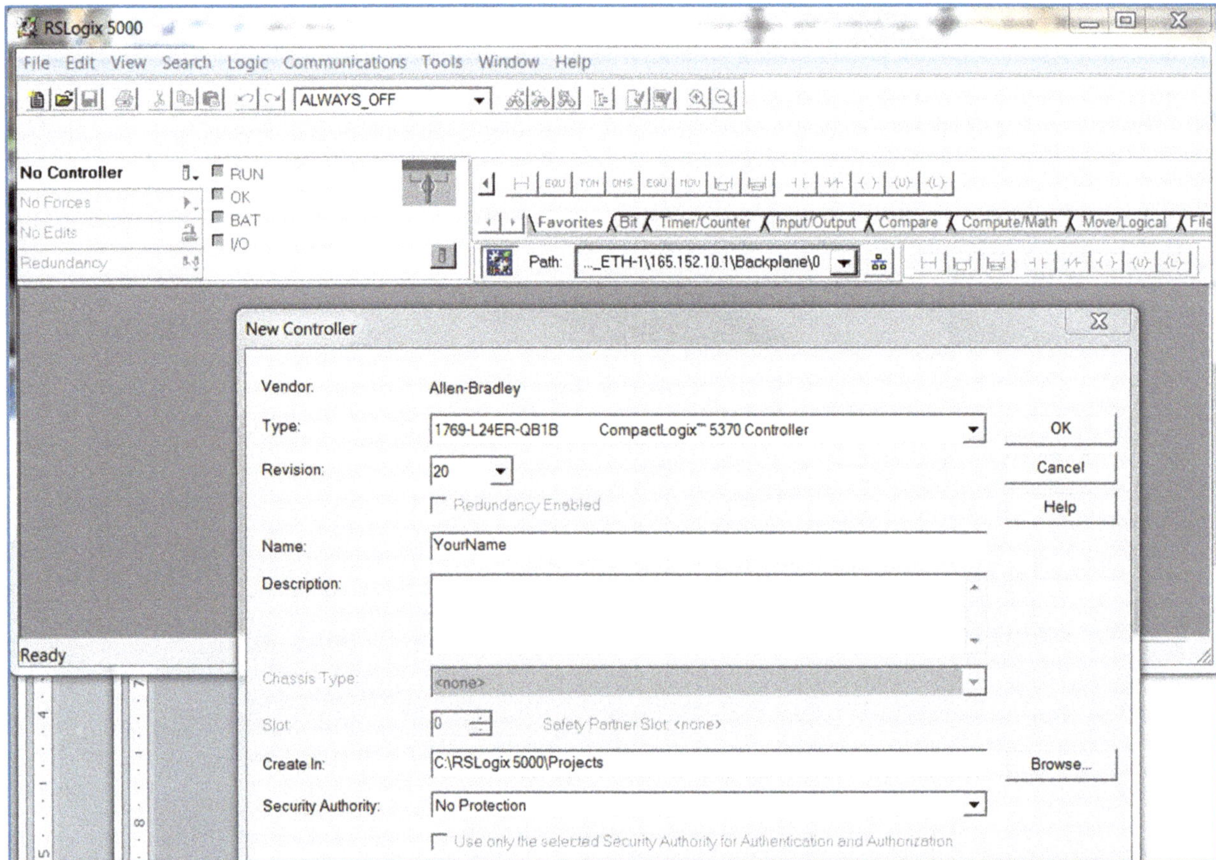

Select the processor and firmware revision and type the name you wish to use. Typically, this will be something descriptive of the machine you are controlling, such as "BearingPress03" or "Line2_Control". For this course, you may wish to use your name.

After the processor name has been entered, press "OK". This is when the project is actually created; the default location for the project is shown above in the RSLogix 5000 folder. If you wish to put it in your own designated location, press the "Browse" button.

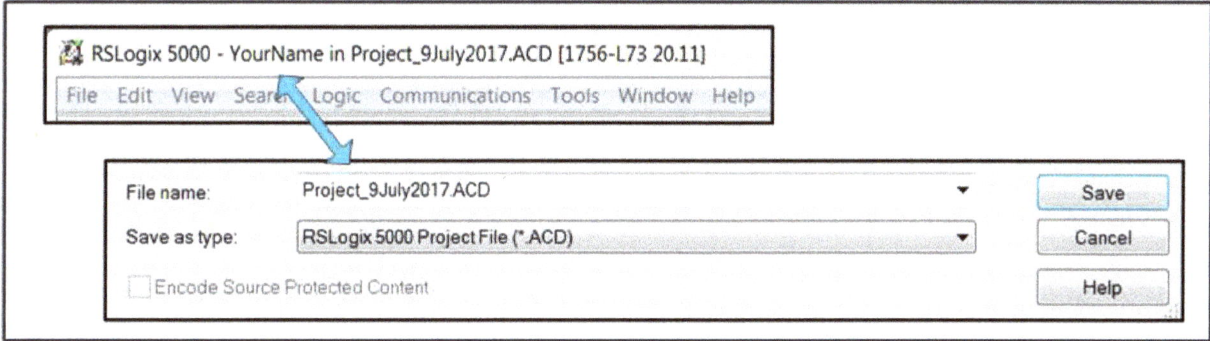

When the project is first created, it will be saved under the processor name. It is also common to save copies of the project with the date embedded, as shown above. When this is done, the header will show the processor name saved "in" the filename.

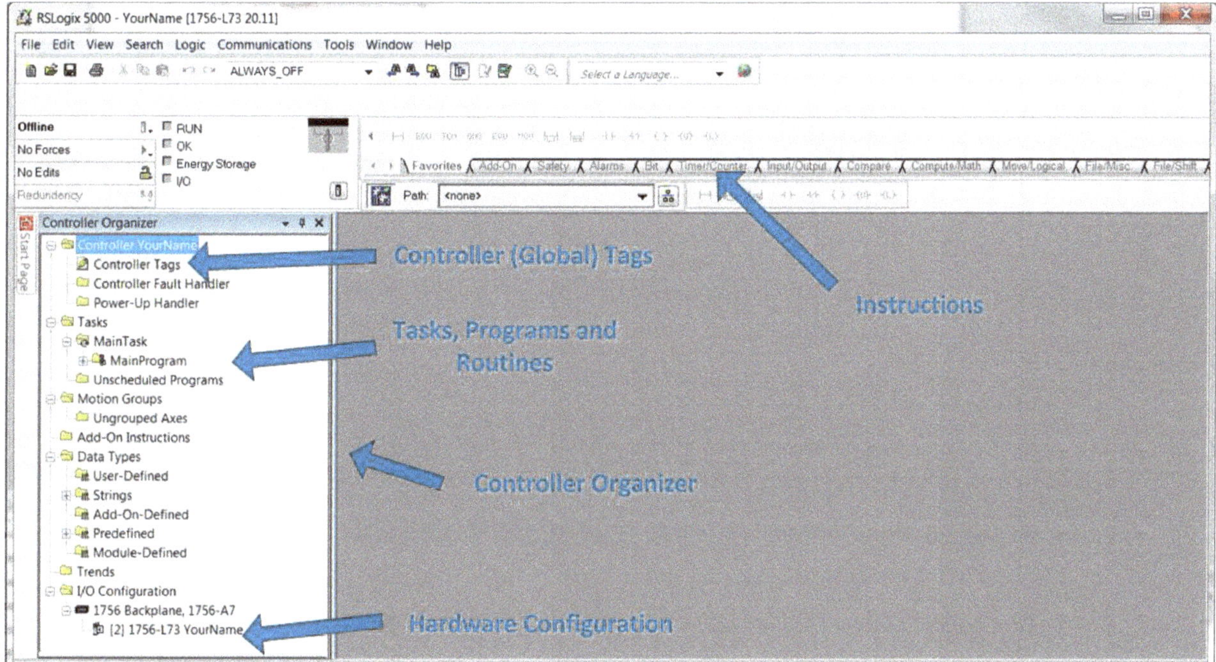

After the project has been created, you are ready to configure hardware, then start designing and programming. The figure above shows some of the different areas of the programming environment. The Controller Organizer on the left side is where you will find all of the different components of your project. This is where you can create new routines and programs, create tags, configure hardware and Ethernet communication networks, and make new data types. Above the Organizer is a status display showing whether or not you are online, in Run mode, or if the processor has a fault.

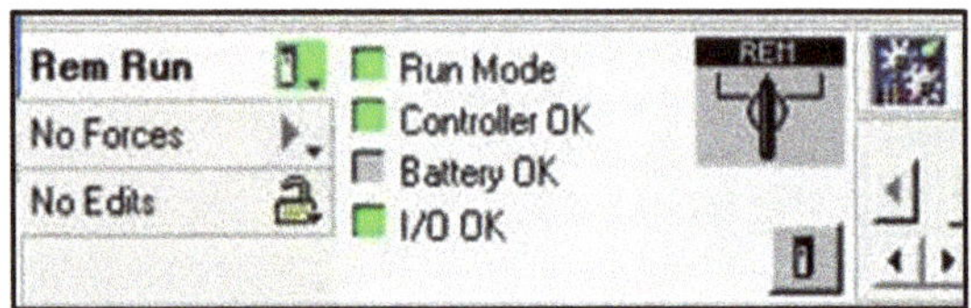

It is important to check the processor status periodically if you are editing online; if editing is done while offline, you will have to download to the processor.

While offline, items in the status area will appear gray; online status will look as shown in the figure above.

Configuring Hardware

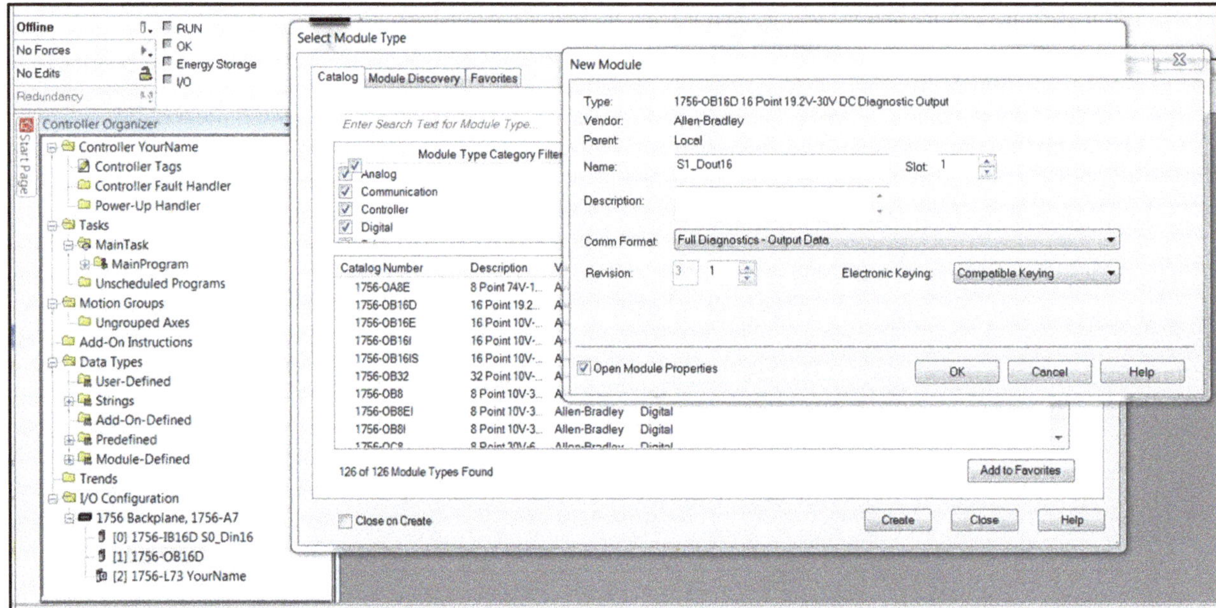

Right-clicking on the Backplane icon will allow selection of modules for your system. Note that in the above figure, the processor was placed in slot 2; the ControlLogix platform allows processors to be placed in any slot. You can even use multiple processors!

Naming your I/O cards is optional. Many different features can be configured for different types of cards; check your documentation for your particular module.

The "Select Module Type" window requires you to select the major revision number. This can't be changed after the card is configured, so it is important to ensure that the correct number is entered. The minor revision number can be changed online or offline after creation.

Electronic keying options include Exact Match, Compatible Keying (shown; the minor revision must match), and Disable Keying. Validated software must select the Exact Match option.

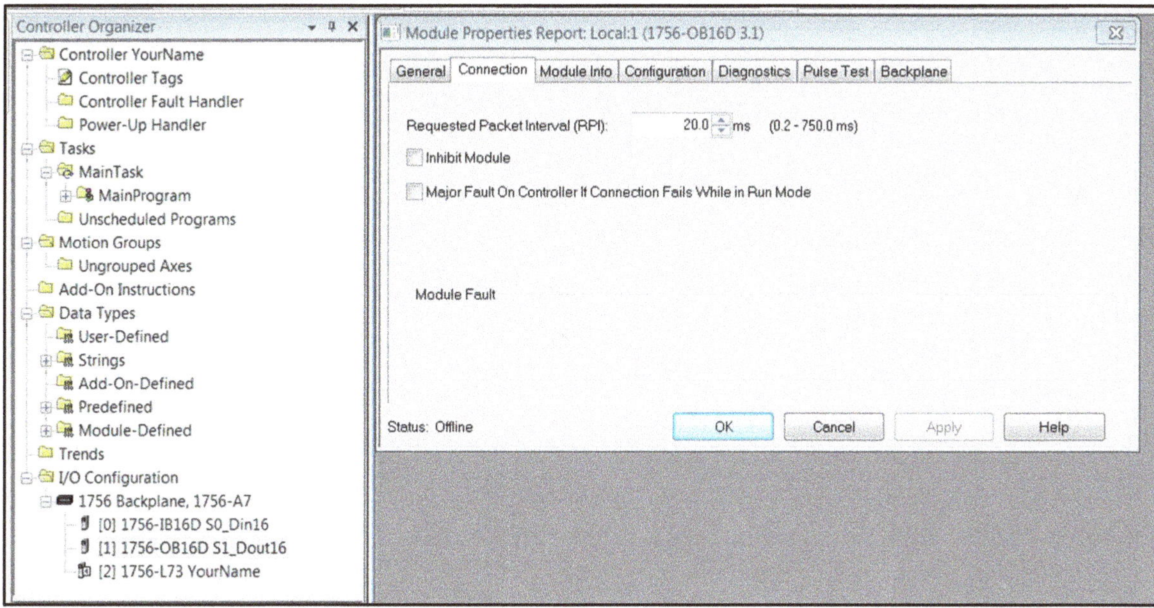

After creating the module, the properties screen will open. There are various settings that can be changed here, but all properties screens have the Requested Packet Interval (RPI) setting. The "Program Processing > Scanning" chapter of PLC Hardware and Programming has an explanation of this setting, as well as a description of how a PLC scans. Take a look at the scanning diagram in the Tasks section of this book.

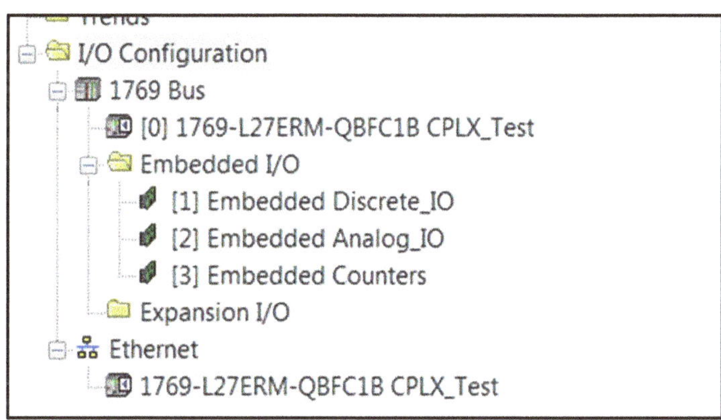

CompactLogix processors often have built-in I/O, and it may not be necessary to add modules. Right-clicking on objects in the I/O configuration and selecting "Properties" will allow them to be configured.

Writing the Program

Tasks, Programs and Routines Overview

After configuring the hardware, programming can begin. Organization of the code is done within the Main Task, which is **Continuous**. There can only be one continuous task, but additional **Periodic** Tasks or **Event** Tasks can also be added. The Continuous Task runs as described in the "Program Processing > Scanning" chapter in this manual.

Tasks can contain multiple programs, which, in turn, contain routines. When the project is created, a "Main Program" is generated, in which is contained a "Main Routine". Programs are added by right-clicking the "Main Task" folder and selecting "New"; routines are added by right-clicking the program.

Each program contains its own Program Tags, which are local to the routines within it. More on this topic is covered in the next section on data.

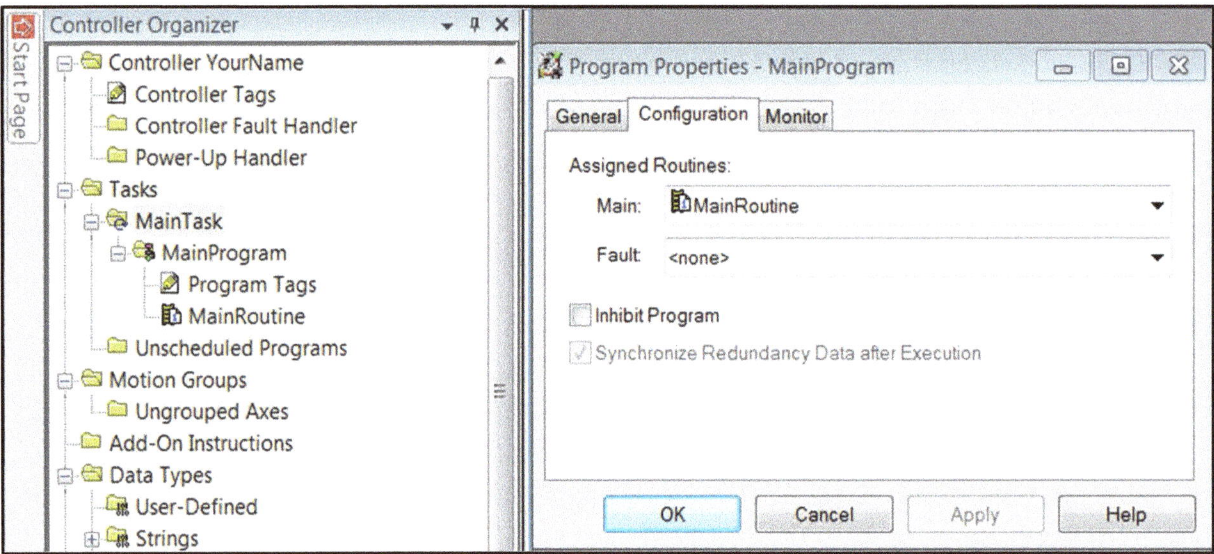

Within each program, a routine needs to be designated to run first. This is done in the dialog box after selecting "Properties" of the program as shown in the figure above. Initially, the "MainRoutine" is configured this way by default, but if you add new programs you will have to select the routine that you want to run as "Main". The name can be anything you wish.

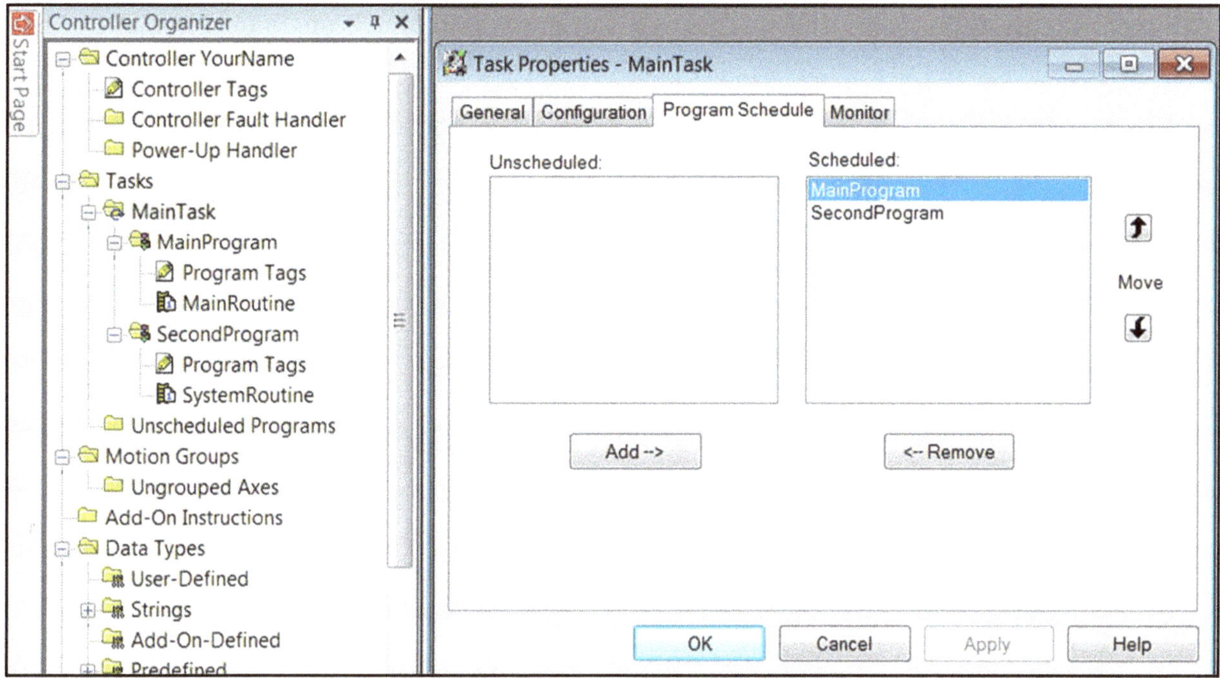

As programs are added to the MainTask, they can be *scheduled* to run in any order. This is done under the Program Schedule tab, which is found under "Properties" of the task. Programs can also be Unscheduled, meaning that they don't run at all.

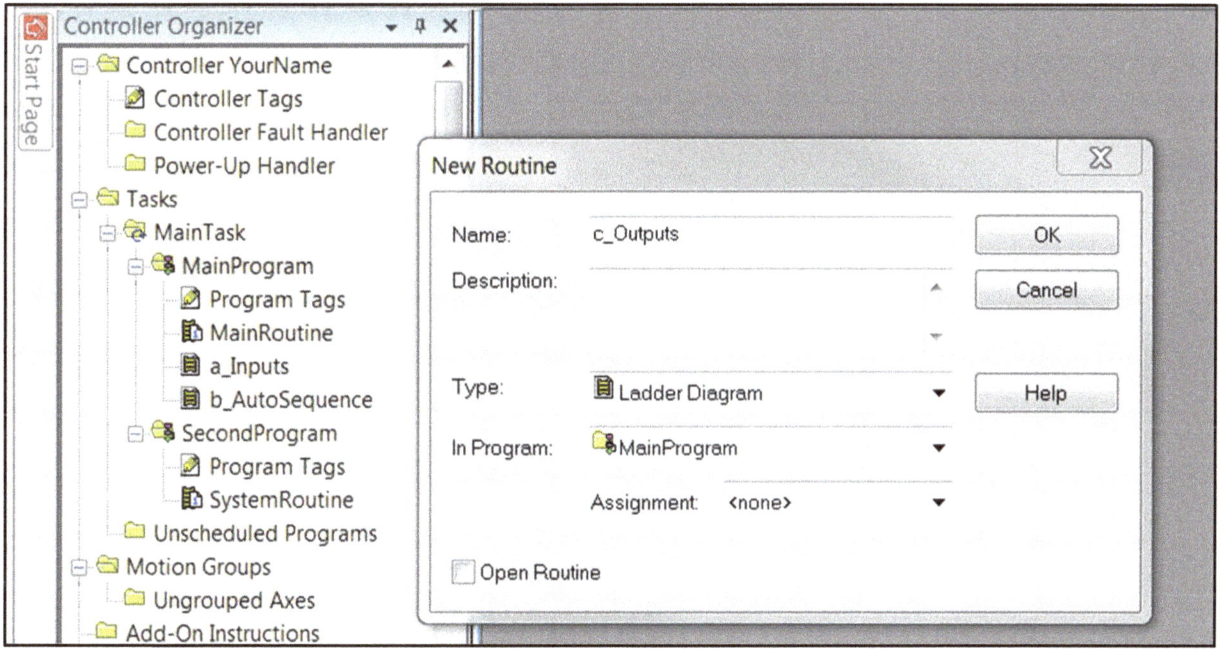

Routines are added by right-clicking the program you want to call the routine from and selecting "New". Routines must be called from the Main routine or the code will not execute. They are not listed in the order they are called; because of this, it is often a good idea to ensure that the routines are alphabetized, as shown in the illustration above, so that the programmer will easily see the order of execution.

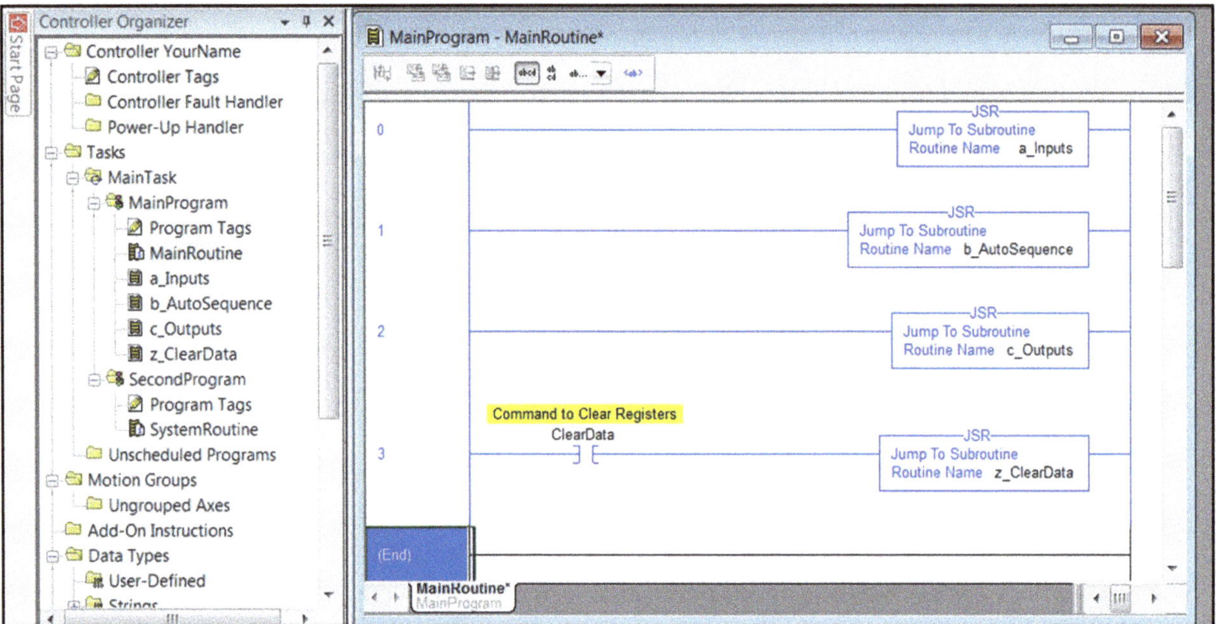

Routines are usually called unconditionally, that is, without a contact in front of the JSR (Jump to Subroutine) instruction. The figure above shows a typical Main routine with unconditional calls, along with a conditional JSR that clears data. Notice that the routines were named in alphabetical order. The MainRoutine and SystemRoutine have a small "1" and appear at the top of the list because they were configured as "Main" in the program properties.

Editing a Routine in Ladder Logic

After a routine has been created, it can be opened for editing by either double-clicking its icon in the Controller Organizer or selecting "Open" after right-clicking on it.

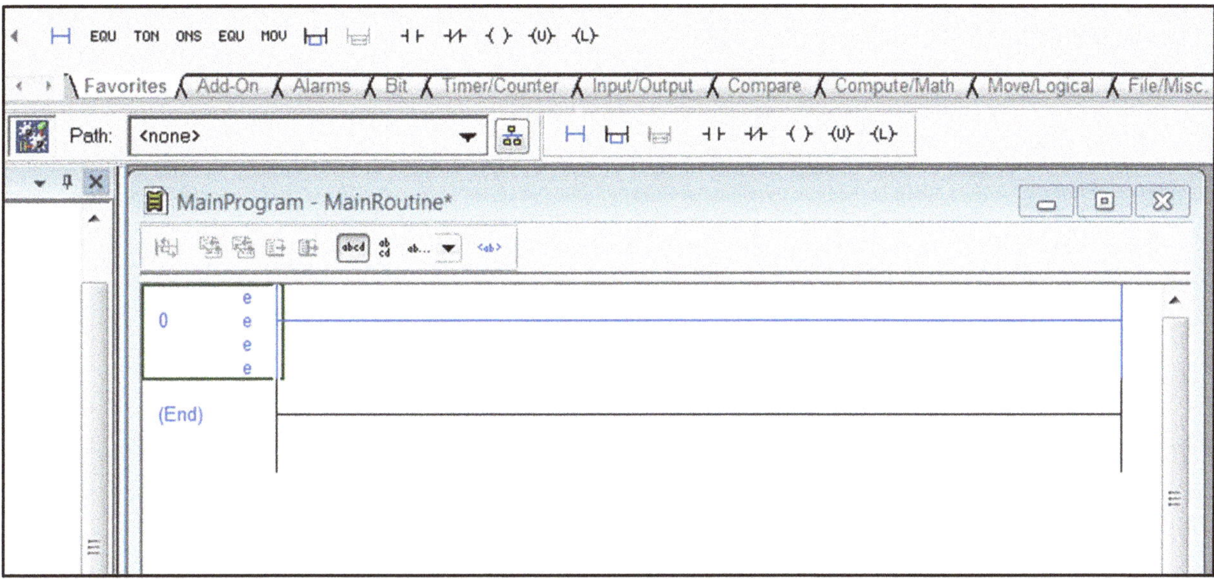

When a new routine is created, it will automatically contain an initial rung of logic as shown above. The lower case "e" means that it is in edit mode; when online, it also signifies that it is not acceptable yet. This can cause problems when new routines are created and not opened before attempting to accept the program or download it.

There are many tabs at the top of the editing window. Selecting one will open a list of **mnemonics** and icons as shown at the upper left of Figure 14. These will differ as each tab is selected.

What is a Mnemonic?

A mnemonic is an abbreviation for an instruction. Allen-Bradley's instructions use three- or four-letter mnemonics to represent instruction names; for example, a normally open contact is abbreviated as XIC, which represents "Examine If Closed". Typing these abbreviations is a shortcut for creating instructions; right-clicking on an instruction and selecting "Change Instruction Type" will allow the programmer to type a mnemonic to replace the instruction.

A list of mnemonics is included in the <u>PLC Hardware</u> section.

After inserting instructions, they must be given an address or name. This is done by creating or using an existing **Tag**.

ControlLogix and CompactLogix Data

Both the ControlLogix and CompactLogix are tag-based systems, supporting all IEC data types. User defined types (UDTs) can also be defined by the programmer. Tags may be up to 40 characters in length and can be composed of alphanumeric characters and the underscore (_) character.

Data Type	Name	Description
BOOL	Boolean	A single on or off element. A bit.
SINT	Single Integer	8 Bits, or a Byte
INT	Integer	16 Bits, or a Word
DINT	Double Integer	32 Bits, or a Double Word
REAL	Real Number, or Float	32 bit Floating Point Number

Many other complex data types are made up of these "elementary" types. Arrays are also supported up to three dimensions.

The ControlLogix platform operates on a "Producer - Consumer" object model, where the Controller and other physical objects such as I/O cards produce and consume information based on a time schedule, known as an RPI, or "Requested Packet Interval". Tags sent and received from other ControlLogix-based systems can also follow this protocol.

Older PLC platforms, such as the PLC 5 and SLC 500, used numeric addresses corresponding to registers. With these PLCs, addresses are categorized by the type of address (bit, integer or real/floating point). Additional registers for the PLC processor or program status were also used. Timers, Counters and control words for more complicated instructions also had their own registers.

Tags can be named however the programmer wishes. Tags still point to memory address registers, but the user doesn't need to know where the registers are. They do need to know what type of data a tag represents though.

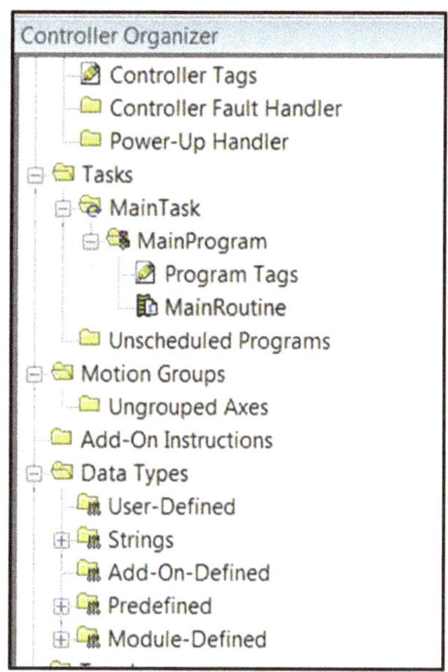

In addition to the simple data types, more complex types, such as timers and counters, also are made up of the data types listed. The figure on the left shows the section of the Controller Organizer that contains all of the different data types; the folder labeled "Predefined" contains all of Allen-Bradley's instruction-based data types. The Module-Defined folder contains all of the data structures for I/O cards, while the Strings folder contains different sized String (text) arrays. User Defined and Add-On-Defined tags are created by the programmer.

Arrays

An Array is a group of similar data types. If 10 Floating Point or REAL numbers are needed, an array can be created as shown below:

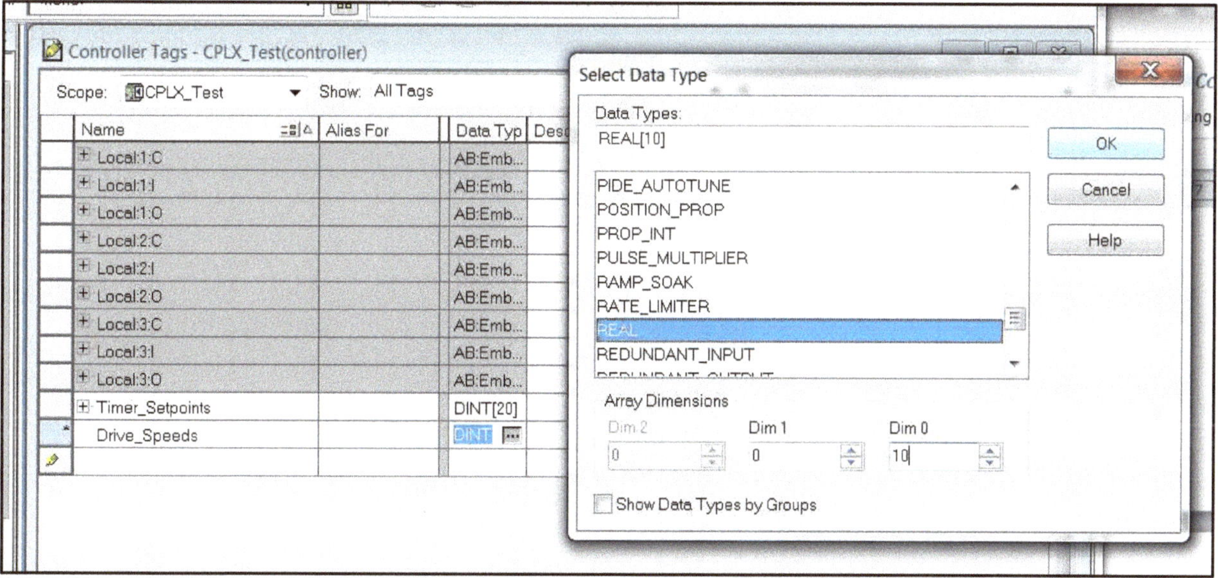

After the name of the tag is typed into the field, the data type is selected by either using a drop-down menu or by typing in the data type. Entering a number or numbers into the dimension fields will allow that number of tags to be created at once. The tag named "Drive_Speeds" will have 10 REAL numbers in it if the OK button is pressed. The tags will be named Drive_Speed[0], Drive_Speed[1] etc., up to Drive_Speed[9].

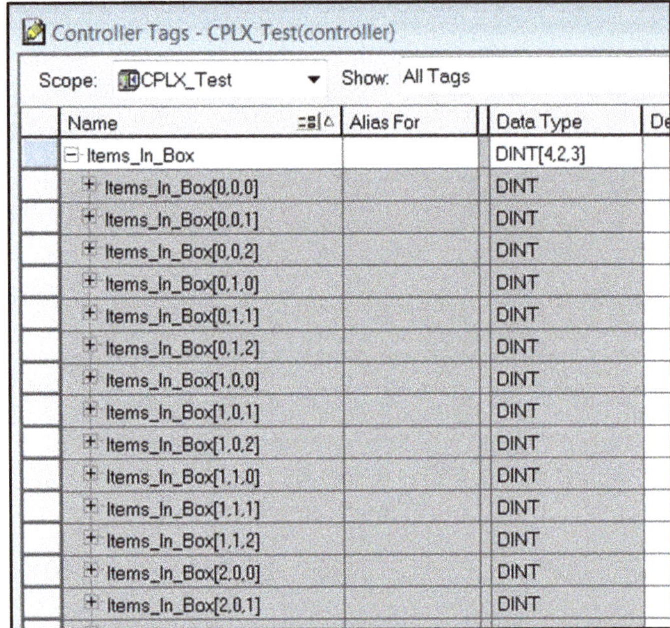

Multi-dimensional arrays can be useful for physical objects. For instance, if it was important to keep track of the number of items in a stack of boxes, a tag could be created with three dimensions, one each for level, column and row.

Of course, it would then be important to remember which name represented each dimension direction!

User Defined Data Type (UDT)

A User Defined Data Type, or UDT, can be used to create groups of data that represent an object, such as a recipe, a manufacturer's product, or a device, such as a Variable Frequency Drive (VFD).

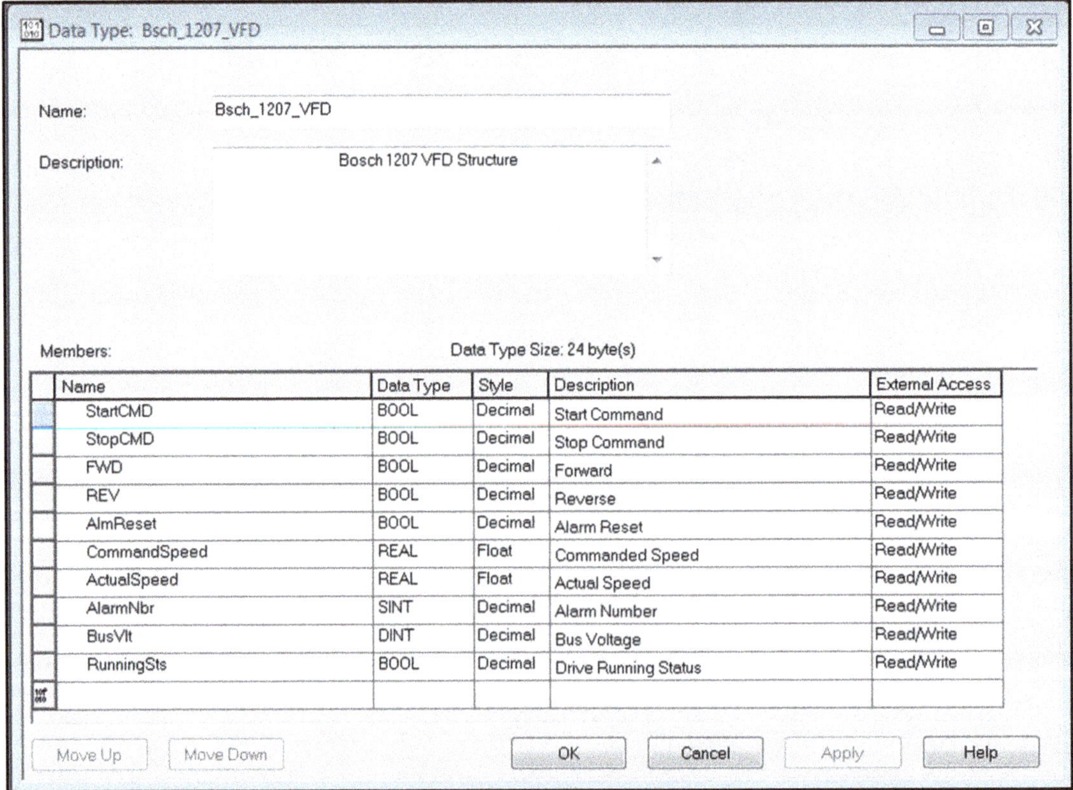

To create a UDT, right-click on the User-Defined folder in the Data Types section of the Controller organizer. Items are typed into the form very similarly to how a new tag is created.

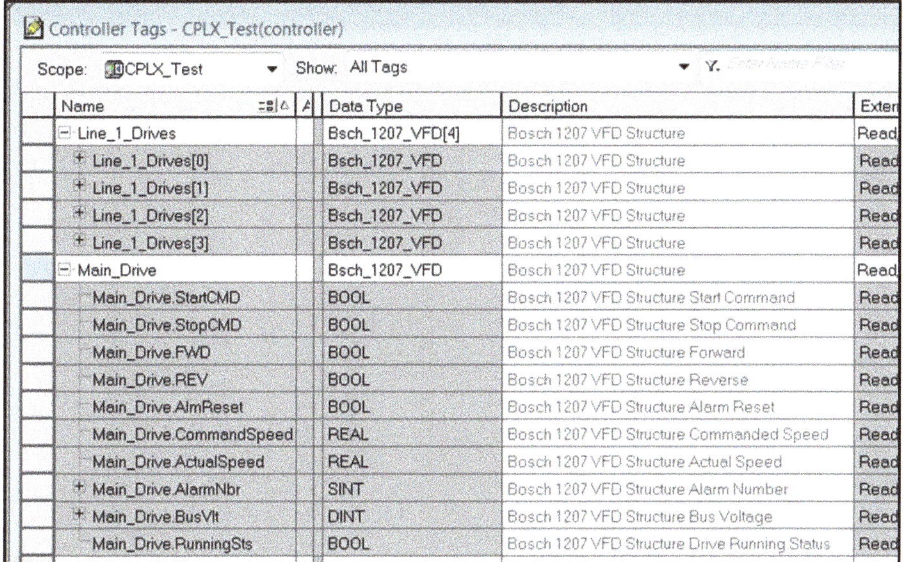

Note in the figure at left that the UDT itself is <u>not a tag</u>; instead, the newly created tag uses the UDT as its data reference. After saving the UDT, its name will be selectable as a data type. The comments are represented in the description, and UDTs can be created as arrays.

Global Tags

Rather than selecting addresses from registers as with the older Allen-Bradley platforms, a database will be created that looks something like the figure below:

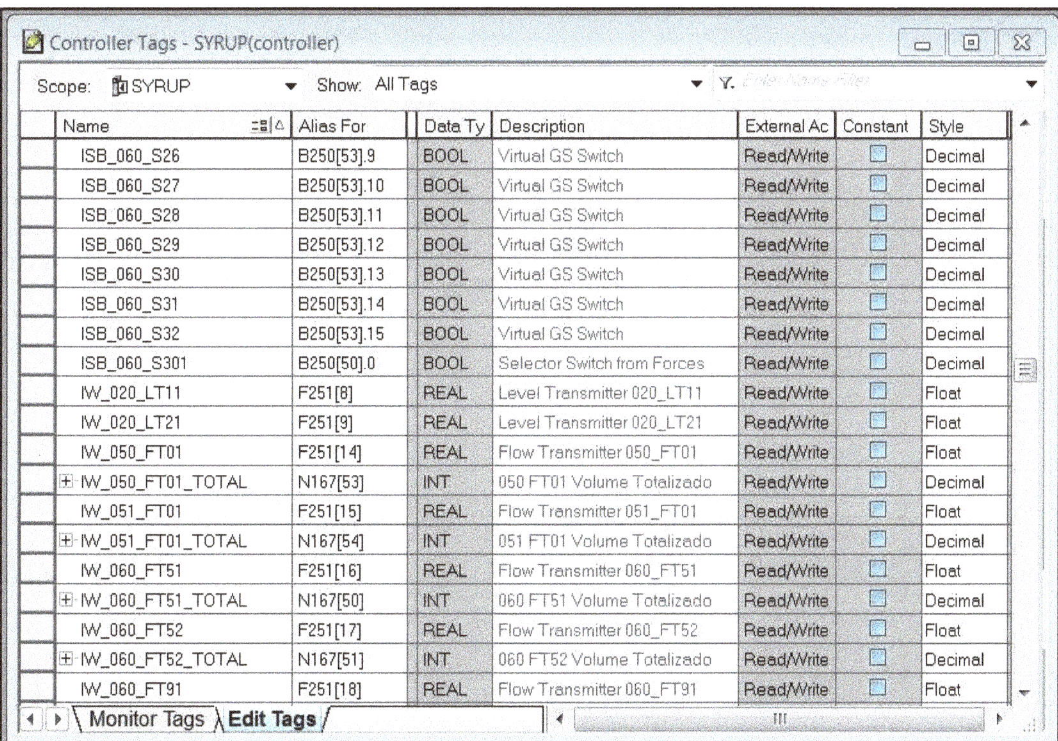

Notice that these tag names are not very descriptive and that they have an "Alias" column with register-type addresses in it. The tag names are derived from P&ID symbols, and the aliases come from another Allen-Bradley PLC platform, RSLogix500. This program was converted from a SLC500 program into this newer ControlLogix processor!

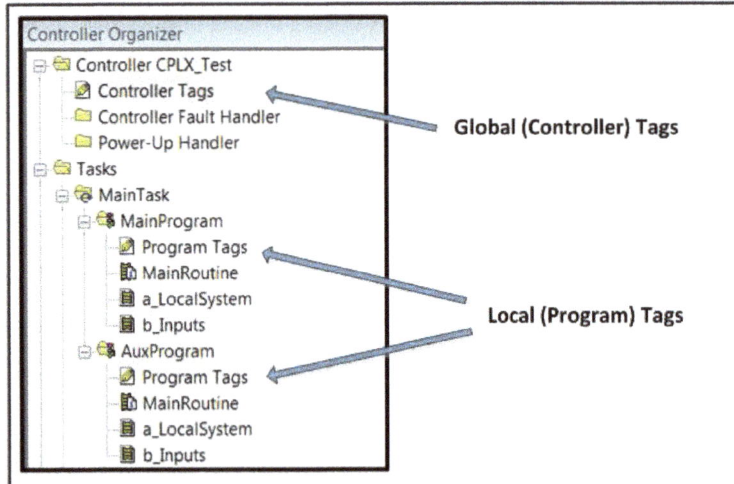

The "Scope" window at the top of the tag database dialog box shows the name SYRUP and has an icon that looks like a processor card. This indicates that the name of the processor is SYRUP and that this list of Controller tags is accessible throughout the program. Controller tags are global, which means that all programs within the PLC can access them.

Program (Local) Tags

Most tag names will be more descriptive than the previous example as shown in the figure below:

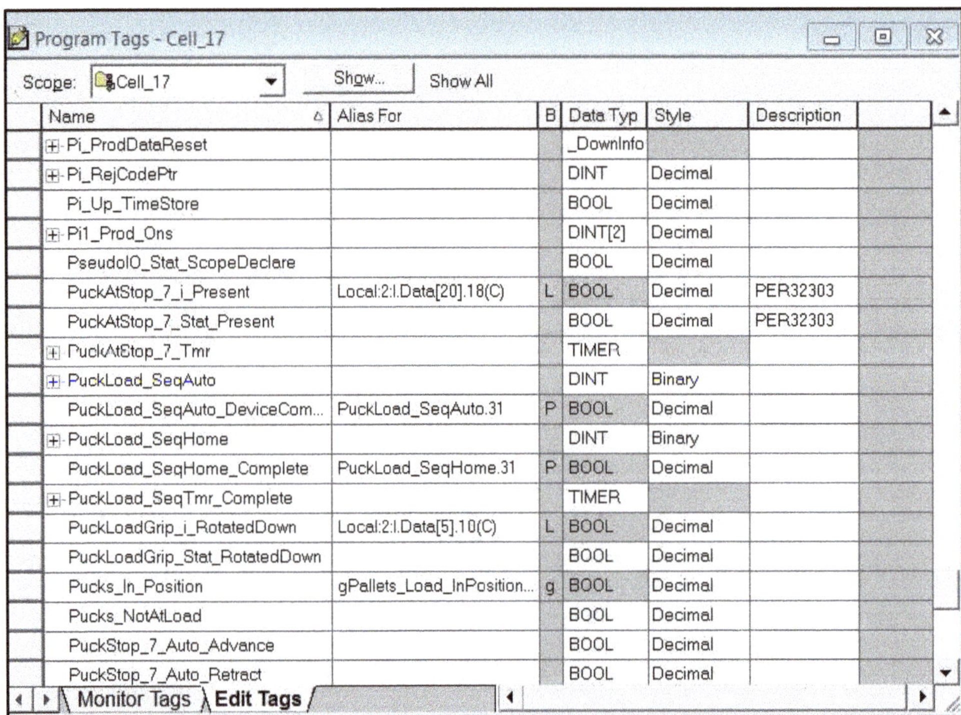

Program Tags - Cell_17

Scope: Cell_17   Show...   Show All

Name	Alias For	B	Data Typ	Style	Description
Pi_ProdDataReset			_DownInfo		
Pi_RejCodePtr			DINT	Decimal	
Pi_Up_TimeStore			BOOL	Decimal	
Pi1_Prod_Ons			DINT[2]	Decimal	
PseudoIO_Stat_ScopeDeclare			BOOL	Decimal	
PuckAtStop_7_i_Present	Local:2:I.Data[20].18(C)	L	BOOL	Decimal	PER32303
PuckAtStop_7_Stat_Present			BOOL	Decimal	PER32303
PuckAtStop_7_Tmr			TIMER		
PuckLoad_SeqAuto			DINT	Binary	
PuckLoad_SeqAuto_DeviceCom...	PuckLoad_SeqAuto.31	P	BOOL	Decimal	
PuckLoad_SeqHome			DINT	Binary	
PuckLoad_SeqHome_Complete	PuckLoad_SeqHome.31	P	BOOL	Decimal	
PuckLoad_SeqTmr_Complete			TIMER		
PuckLoadGrip_i_RotatedDown	Local:2:I.Data[5].10(C)	L	BOOL	Decimal	
PuckLoadGrip_Stat_RotatedDown			BOOL	Decimal	
Pucks_In_Position	gPallets_Load_InPosition...	g	BOOL	Decimal	
Pucks_NotAtLoad			BOOL	Decimal	
PuckStop_7_Auto_Advance			BOOL	Decimal	
PuckStop_7_Auto_Retract			BOOL	Decimal	

Monitor Tags \ Edit Tags

Program tags can only be seen or accessed by the particular program they are in. This particular program has the name "Cell 17"; other programs within the controller, such as those from the program "Cell 16", cannot access these tags.

Aliases

An alias is a "nickname" for a tag. In the previous figure, the tag "PuckAtStop_7_i_Present" has an address listed next to it of Local:2:I.Data[20].18. This means that the tag is connected to the physical device at that local address. Since the address is Local:2, it is accessed through slot 2 of the local I/O rack. In this case, that card is a DeviceNet card with nodes assigned to it, and Node 20 is the place that the tag is linked to. If the physical input changes state, the corresponding aliased address will do the same.

It is common to alias physical inputs and outputs to more descriptive tag names like this. To alias a tag to another, simply browse for the tag you wish to alias to in the "Alias For" column.

Add-On Instructions (AOIs)

In addition to the instructions provided by Allen-Bradley and listed in this book, it is also possible to build custom instructions.

To create an AOI, click inside the Add-On Instructions folder in the Controller Organizer and select New Add-On Instruction. A dialog box will open as shown below:

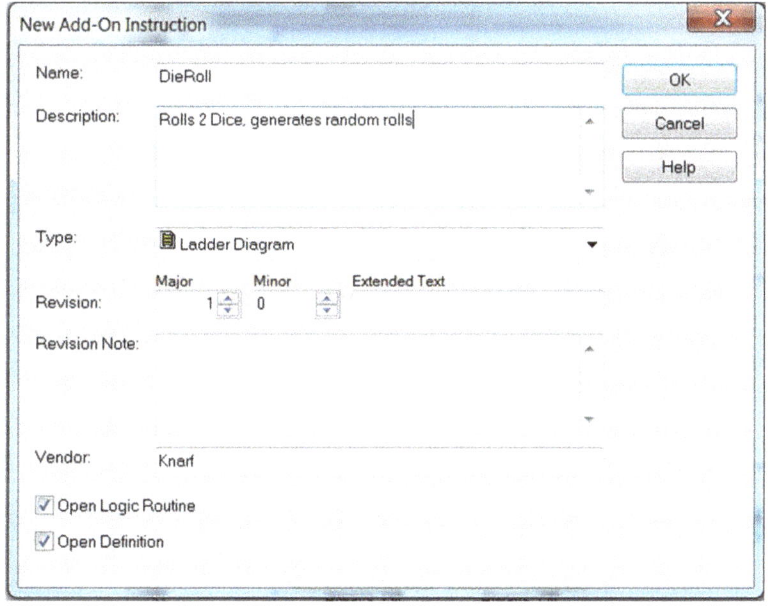

The purpose of an AOI is to create a reusable instruction that can be used in multiple projects, just like any other instruction in the tabs at the top of the editor.

This example is a fun instruction that acts as a random number generator: it rolls a set of dice.

The name of the instruction will appear as whatever you place in the Name field, so shorter names are better.

Since the boxes at the bottom of the New Add-On Instruction dialog are checked, both the logic routine and the Definition dialog will open.

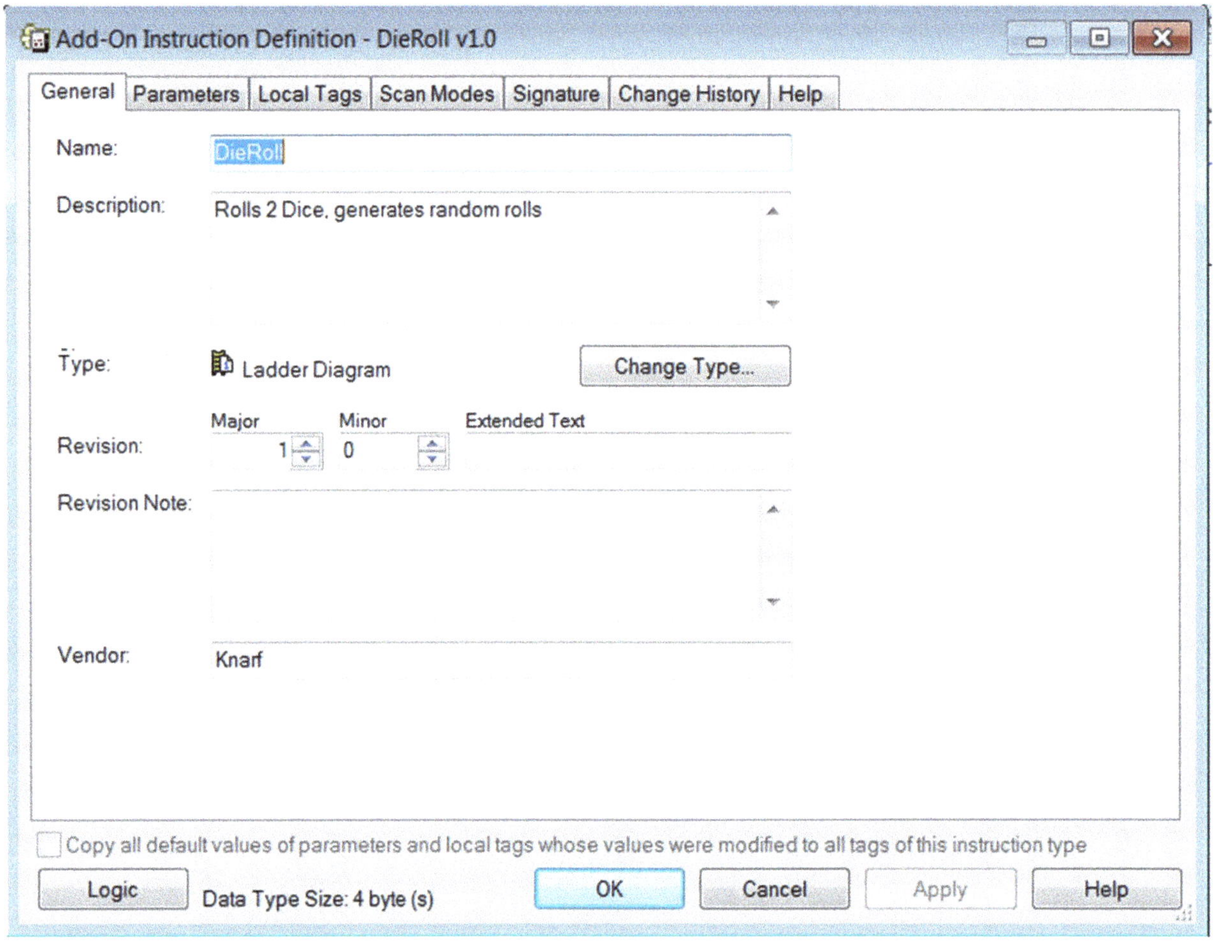

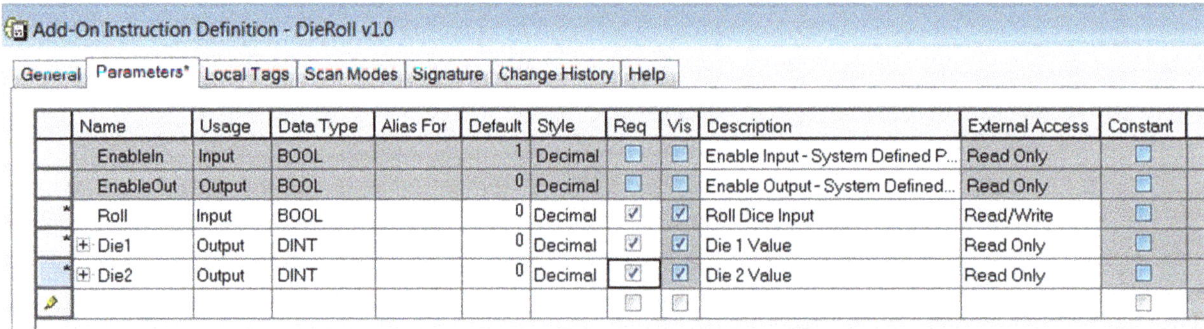

The Definition can be opened by right-clicking the name of the instruction in the Add-On Instructions folder and selecting "Open Definition" at any time. It contains a number of tabs that can be selected to configure various features of the instruction. The General tab is shown above.

Add-On Instruction Definition - DieRoll v1.0

General | Parameters* | Local Tags | Scan Modes | Signature | Change History | Help

	Name	Usage	Data Type	Alias For	Default	Style	Req	Vis	Description	External Access	Constant
	EnableIn	Input	BOOL		1	Decimal	☐	☐	Enable Input - System Defined P...	Read Only	☐
	EnableOut	Output	BOOL		0	Decimal	☐	☐	Enable Output - System Defined...	Read Only	☐
*	Roll	Input	BOOL		0	Decimal	☑	☑	Roll Dice Input	Read/Write	☐
*	± Die1	Output	DINT		0	Decimal	☑	☑	Die 1 Value	Read Only	☐
*	± Die2	Output	DINT		0	Decimal	☑	☑	Die 2 Value	Read Only	☐
							☐	☐			☐

The **Parameters** tab is used to configure the Input and Output variables that will be used to pass data into and out of the instruction. The Name will appear inside of the instruction if the Vis (Visibility) box is checked for that row, and the parameter is optional if the Req (Required) box is not checked.

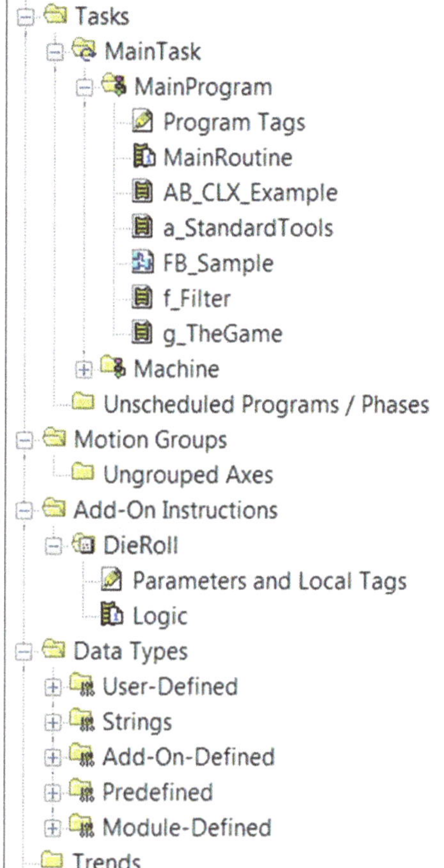

After saving the configuration, the AOI appears in the folder as shown to the left. As mentioned previously, double-clicking the object will open it for editing.

An important thing to remember when creating AOIs: they cannot be created while online! They can, however, be created in a separate offline project and then imported by right-clicking on the Add-On Instruction folder and selecting "Import Add-On Instruction".

The **Local Tags** tab will contain any internal tags that are created while building the logic. They can be filled in from the tab or be created by right-clicking in the tag field when it is entered in the logic. Double-clicking the Parameters and Local Tags icon opens this also.

The **Scan Modes** tab is used to add additional routines that will run before the Logic routine (Prescan), after the Logic Routine (Postscan) or if the EnableIn parameter is false (EnableInFalse).

The **Signature** tab is used to generate a code that uniquely identifies the instruction and seals it from modifications.

The **Change History** tab logs the time and date of modifications, and the **Help** tab allows the programmer to write descriptive information that documents the instruction and provides information to the end-user.

Roll the Dice!

The logic in this example uses two free running timers to generate numbers based on when the player presses and releases a pushbutton. Since the timer setpoints are both multiples of six, it is easy to divide the accumulator values and have an equal random chance of rolling 1-6.

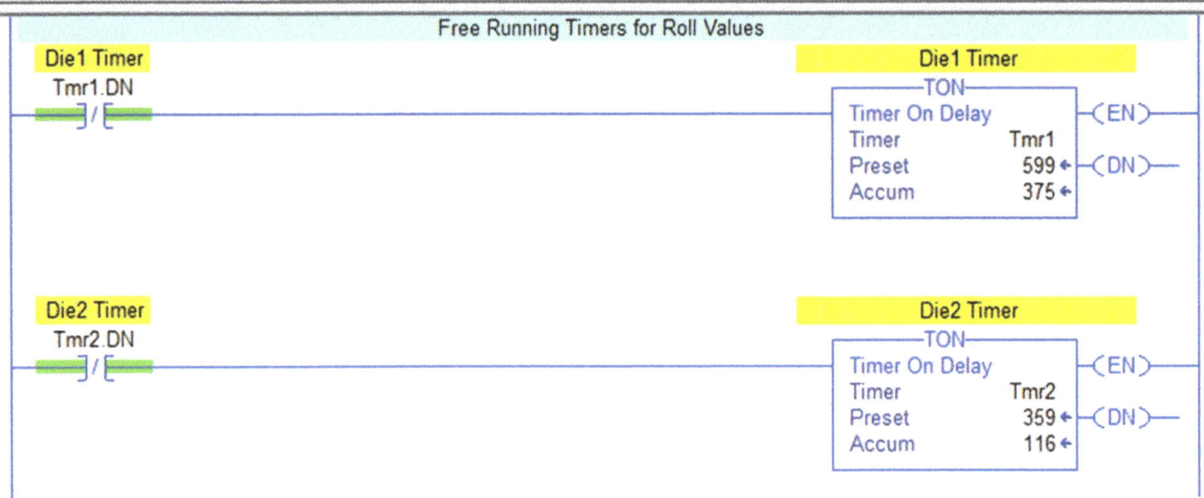

The two free running timers have different setpoints so that they run at different speeds.

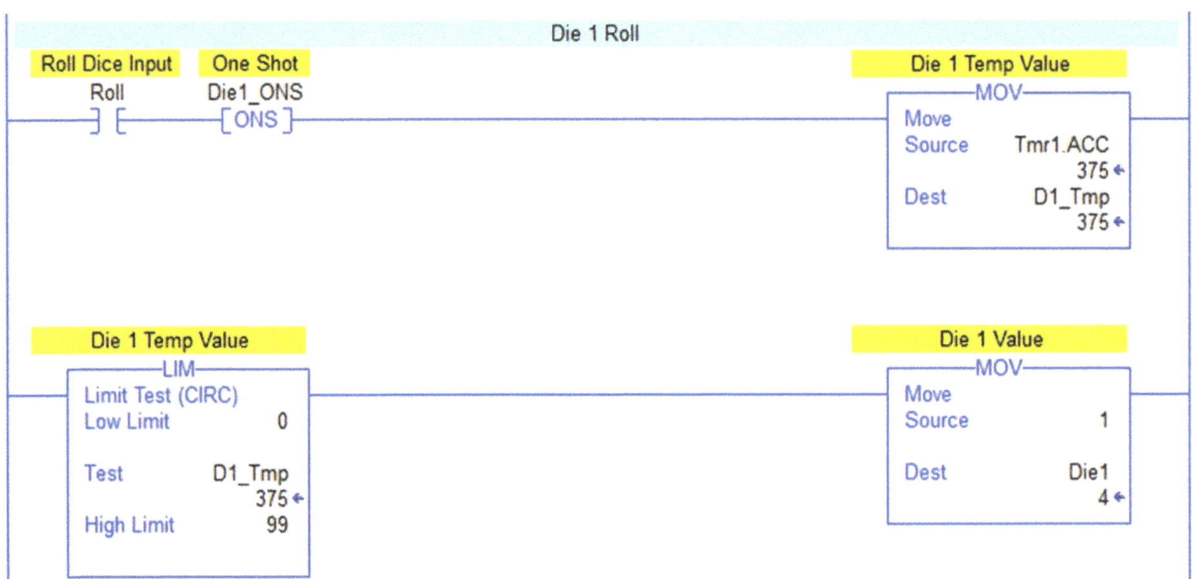

When the Roll input occurs, a one-shot captures the timer accumulator into a temporary variable. A Limit instruction is used to divide the captured value into six different ranges.

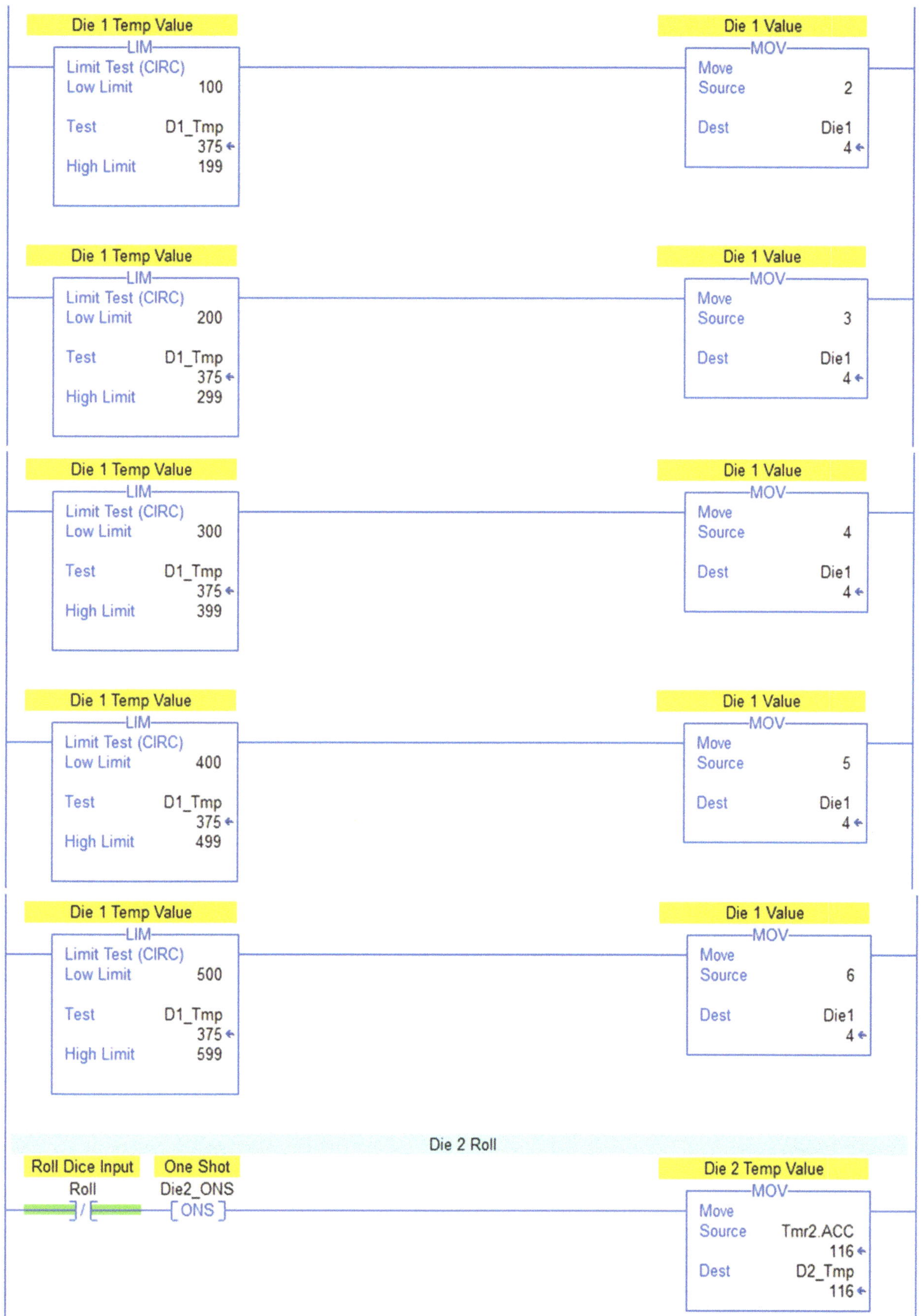

Die 1 Temp Value
LIM
Limit Test (CIRC)
Low Limit   100
Test   D1_Tmp
375
High Limit   199

Die 1 Value
MOV
Move
Source   2
Dest   Die1
4

Die 1 Temp Value
LIM
Limit Test (CIRC)
Low Limit   200
Test   D1_Tmp
375
High Limit   299

Die 1 Value
MOV
Move
Source   3
Dest   Die1
4

Die 1 Temp Value
LIM
Limit Test (CIRC)
Low Limit   300
Test   D1_Tmp
375
High Limit   399

Die 1 Value
MOV
Move
Source   4
Dest   Die1
4

Die 1 Temp Value
LIM
Limit Test (CIRC)
Low Limit   400
Test   D1_Tmp
375
High Limit   499

Die 1 Value
MOV
Move
Source   5
Dest   Die1
4

Die 1 Temp Value
LIM
Limit Test (CIRC)
Low Limit   500
Test   D1_Tmp
375
High Limit   599

Die 1 Value
MOV
Move
Source   6
Dest   Die1
4

Die 2 Roll

Roll Dice Input   One Shot
Roll   Die2_ONS
]/[[ONS]

Die 2 Temp Value
MOV
Move
Source   Tmr2.ACC
116
Dest   D2_Tmp
116

Die 2 is captured with a one-shot when the player releases the button.

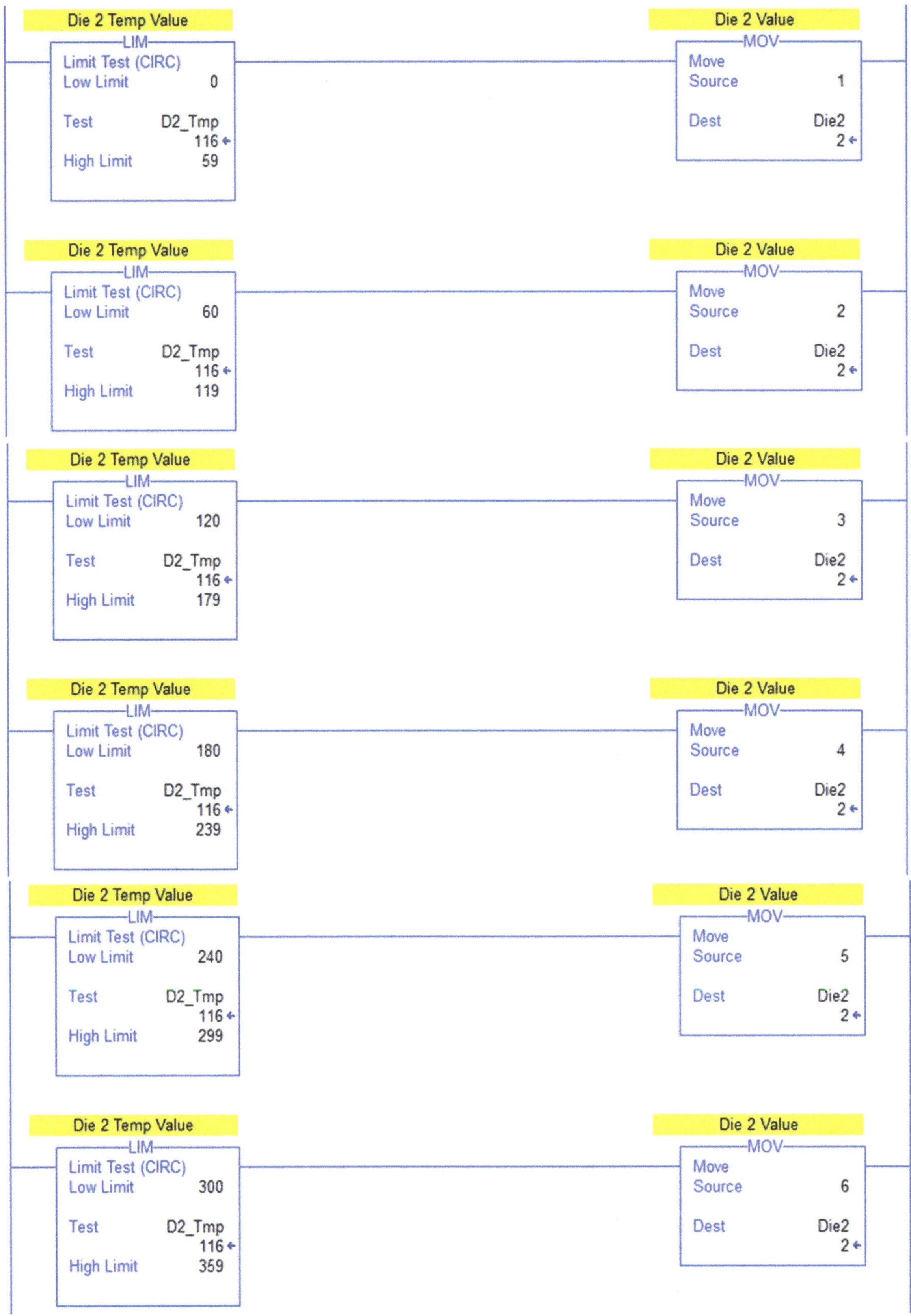

Die 2 ranges are 60 instead of 100.

After the logic is written, the AOI is ready to use. It appears in the Add-On folder as DieRoll since that was the name it was created under. When it is placed into a Rung, it appears as in the diagram below.

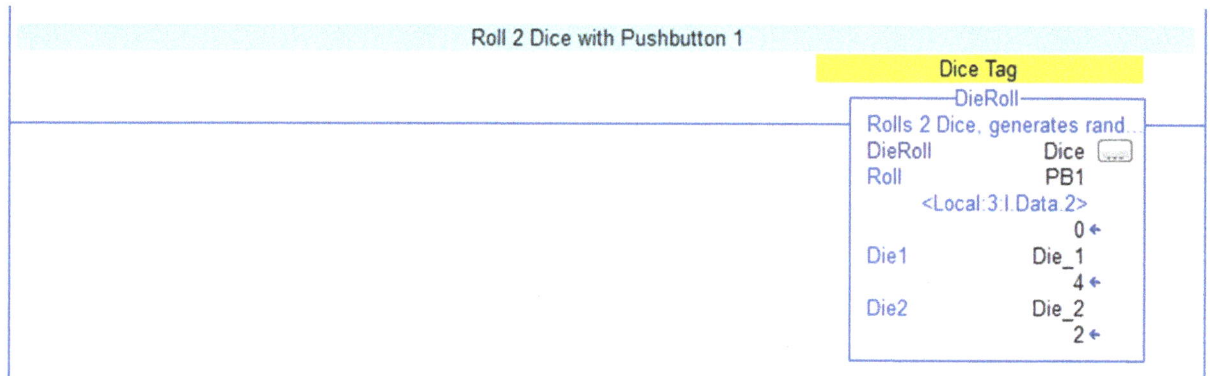

The AOI requires a tagname for each use of the instruction; in this case, it is "Dice". The tag includes the input and output variables created under the Parameters tab, as well as the EnableIn and EnableOut BOOLs.

The input assigned is a pushbutton aliased to Input 3.2 in an Allen-Bradley ControlLogix PLC, and the Die_1 and Die_2 tags are integer values.

Add-On Instructions are a powerful tool and can be used when there is a need for reusable code or it might be used in different processors. Programmers may create a library of AOIs that they use in their projects, or companies that build standardized equipment might use them for common tasks. It is also a convenient method of locking code so that other programmers can't change it or even see how it works.

Other Languages

In addition to Ladder Logic, which is the main language used by Allen-Bradley programmers, the other IEC 61131 languages are also supported to varying degrees.

ASCII Mnemonics (Instruction List)

While it is possible to view ladder logic as text, it is not common to use this feature other than for importing and exporting logic.

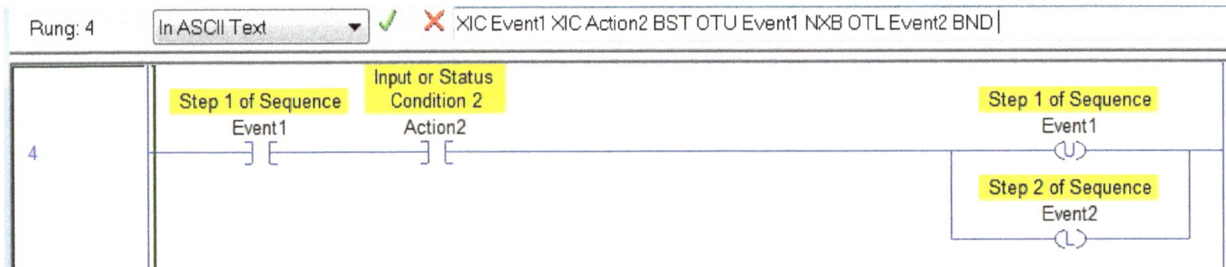

Double-clicking to the left of the rung shows the field at the top of the editor. The text description for this rung is (XIC (Examine If Closed, Normally Open Contact)) <Tagname> etc. The BST is Branch Start, the OTU and OTL are Unlatch and Latch, and the NXB is Next Branch.

Spreadsheets are often used to concatenate strings like this, substituting tags to duplicate rungs. A sample rung can first be exported as a CSV or XML file and opened in Microsoft Excel. The rung can then be disassembled and different tags substituted, making the chore of duplicating hundreds of duplicate rungs with different tags (Such as Faults, Input and Output rungs) much less tedious.

Function Block Diagram (FBD)

The programming environment can be selected as FBD when creating a new routine. FBD is part of the Standard and Lite (CompactLogix) versions of software.

Structured Text (ST)

Structured Text is not part of the standard programming software, but it can be purchased as a separate module. It is included in the Full and Professional versions.

Sequential Function Charts (SFC)

As with Structured Text, SFC can be purchased as a separate module or is included in the Full and Professional versions.

RSLogix5000 - Program Structure and Advanced Topics

Program organization is in Allen-Bradley is done primarily so that programmers can easily find areas of interest in the program. It is also a best practice in Object Oriented Programming (OOP).

Project organizational components include Tasks, Programs and Routines.

Tasks

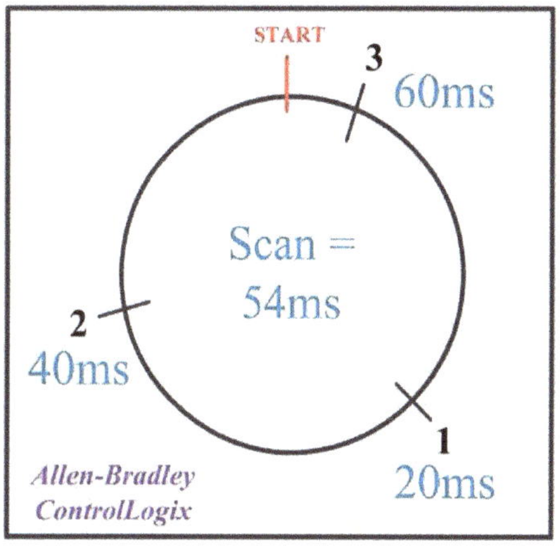

All PLCs have at least one **Continuous** task that runs as fast as the CPU allows based on the number of instructions in the task. The basic PLC Hardware and Programming book describes how the ControlLogix platform scans logic; rather than updating the Input and Output tables at the beginning and end of the scan, I/O is updated according to hardware settings on each card, this is the **Requested Packet Interval**, or **RPI**.

Because Allen-Bradley rungs are scanned until the logic is false, scan times can vary quite a bit. Some platforms scan all of the logic to the end of each rung, but not Allen-Bradley. When the rung loses continuity, the scan moves to the next rung.

It might seem as if this makes I/O in the continuous scan update more quickly, but it still takes the same amount of time for the rung evaluation to proceed through the scan. Though the inputs and outputs are updated three times in the diagram above, it still takes 54 milliseconds for the task to make its way through the logic and activate the outputs.

The Logix 5000 platform has two additional types of task in addition to its one continuous task.

The **Periodic** task can ensure that the logic is evaluated faster than the continuous task. Periodic tasks are set to run at a higher priority than the continuous task, so if logic with output rungs is placed in a periodic task with a 10ms setting, it will be sure to update the outputs at least at the 20ms period shown above. Of course, if there is more code than can be evaluated in the 10ms time allotted, it will take longer.

Periodic tasks are often set to execute at a longer interval than a typical continuous scan. One instance is when running PID loops on analog I/O, a typical setpoint for this interval is 100ms. Another case is when collecting production data, this period may be set at a period of one second or more.

As mentioned previously, priority levels can be set for these tasks; a lower number indicates a higher priority. This can be used when periodic task values are multiples of each other and would occur at the same time; for instance, a 100ms task and a 500ms task will overlap every 5 cycles of the 100ms task.

Event tasks are tasks that run based on an external signal. This can be a physical input if the Change of State selection is made on an input card, an Event instruction placed in the scan, or various motion control Axis events. Produced tags in a remote PLC can be assigned to trigger the task also (consumed).

Programs

A Program in the ControlLogix platform allows local tags to be assigned that are only accessible by routines within the program. A Main routine must be assigned within each program to call its routines.

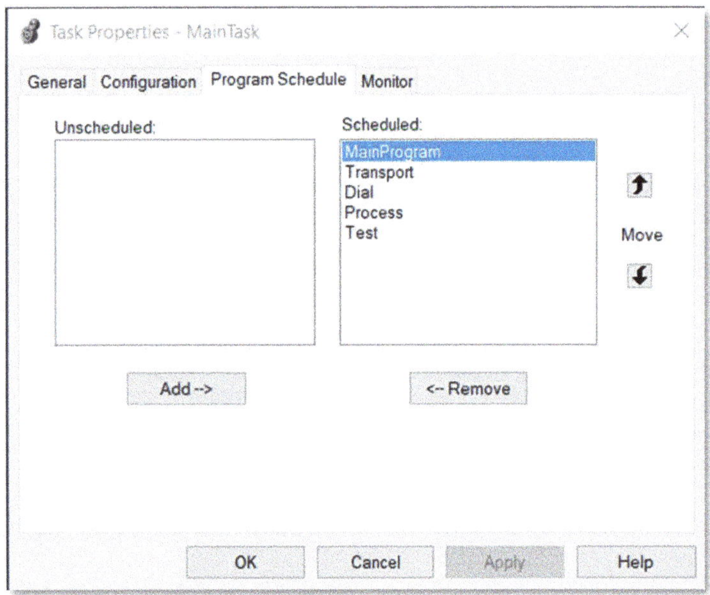

Unlike routines, programs are not alphabetized. They are called by using a Program Schedule, accessed by right clicking the task and selecting Task Properties. Clicking the arrows will move the program up or down in order, and programs can be Unscheduled by clicking Remove. The program will not run at all in this case. This is also a necessary step if the program is to be deleted.

Routines

As mentioned in the programming example, routines are used to organize code into functional sections. Names appear in the Controller Organizer in alphabetical order, so a letter or other designator such as "r01, r02…" is often prepended to the name. Routines may also be used to contain reusable functions and called multiple times.

Structure

Despite its importance, there are no rigid, set-in-stone rules for PLC programming organization. The organizational method described below is one that is commonly used by large machine builders; however, there are other techniques you can choose to employ. The key objective in program organization is to make a machine easy to troubleshoot and modify.

After gathering information about the system to be programmed, the organization and layout of the program itself can begin. There are two ways to separate the sections of the program. The first is by the physical areas of the machinery, and the second is by the type of code. Small mechanical or logical sections of a machine, such as a pick and place, conveyor or dial table will be referred to as "stations". Stations can then be grouped into larger groups.

Process control programs are often divided into functional areas, such as tanks or lines. Process control stations can often be operated independently in Manual or "Hand" operation.

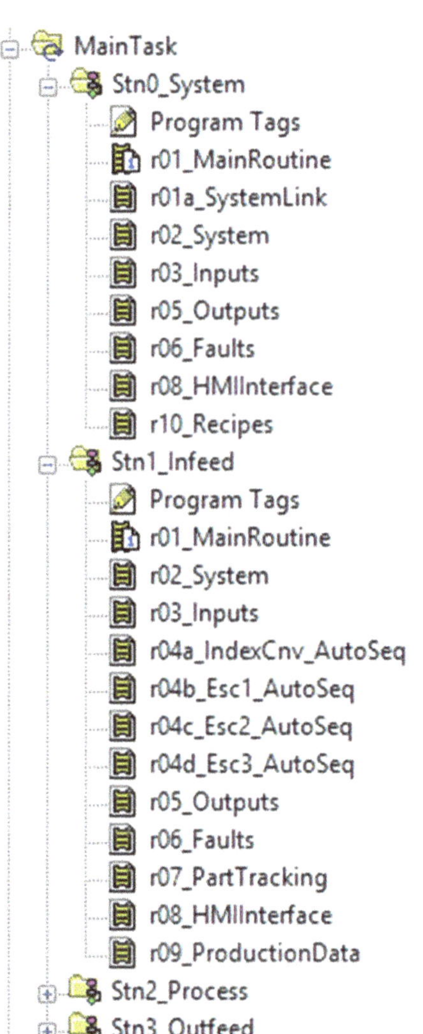

Zones can be used as a logical division of groups of stations. All the mechanism-operating auto sequences can be placed within the station groups. The primary reason for dividing a machine into zones is to allow the zones, or groups of stations, to act autonomously. In this way, zones can be thought of like separate machines. Zones will often each have their own safety circuits so if a guard door opens, for example, it only shuts down the group of stations in that zone. Subsequently, there may be routines within the group called Zone_1, Zone_2, and so forth. Within the groups may be Input, Output and Fault routines, as well as routines that link to all of the stations assigned to the zone. Data acquisition and productivity information may also be assembled at the zone level. Generally, the inputs and outputs would only be those that affect all of the stations, such as safety or indication I/O. A zone may also have an HMI that only controls devices within the zone; HMI bit mapping could also then be contained in the Zone program and routines.

If a platform allows for multiple programs, each with its own group of routines, stations can each have their own program containing standardized routines for system functions, inputs, auto sequences, outputs, and faults or alarms. An example of this from Allen-Bradley's RSLogix 5000 platform is shown in the organizer image on the previous page. Each program contains tags or addresses that are local to the program.

Common Routine Types

System functions include modes of operation and machine states. These are usually modes that describe the condition of the whole machine or zones. Auto Mode, Manual Mode and Auto Cycle are common examples of this. A System Routine will usually contain control logic for these types of functions, along with overall system "housekeeping".

Auto Mode: This mode is an idle state that is a prerequisite for placing a fully-automated machine into Auto Cycle. It usually eliminates the ability to operate individual actuators by pushbuttons or switches, but rather allows a machine to sequence through automated motions.

Manual, Maintenance or "Hand" Mode: This allows actuators and devices to be actuated by means of switches or pushbuttons, outside of the automatic functions. In process control operation, devices are often individually able to be placed in "Hand" mode, even while other devices or systems are being controlled automatically.

Auto Cycle: This mode is a subset of Auto Mode. A machine is usually placed into Auto Cycle by holding a button for a period of time while a warning signal sounds. This allows personnel in the immediate area to stop the operator from starting the machine's operation if a hazard is present. The warning signal often pulses or "beeps".

Other requirements, such as only allowing Auto Cycle start when a machine is at a home or origin position, only allowing the progression of an auto sequence when the machine is in Auto Cycle, and only allowing cycle stop when the machine or sequences are in a specific position or condition, are common.

Other optional modes for machinery:

Dry Cycle: Allows machine to run in Auto Cycle with no parts. Often used to "exercise" a machine during Factory Acceptance Testing (FAT) or Site Acceptance Tests (SAT).

Single Cycle: Allows machines to process parts one at a time in Auto Cycle or run automatic sequences, such as pick and places, in Manual Mode.

Single Step: Allows automatic sequences to run one step at a time. Typically only used for debug. Requires a single-step button in addition to mode control enable/disable.

Purge or Runout: Allows a machine to empty without bringing new parts into the system. Usually used during product changeover.

Homing: Allows actuators to return to their starting or origin location automatically. May be done in Auto or Manual mode depending on preference. Typically, a sequence is used to ensure that actuators are moved in the correct order; see the Auto Sequence section for how to write this.

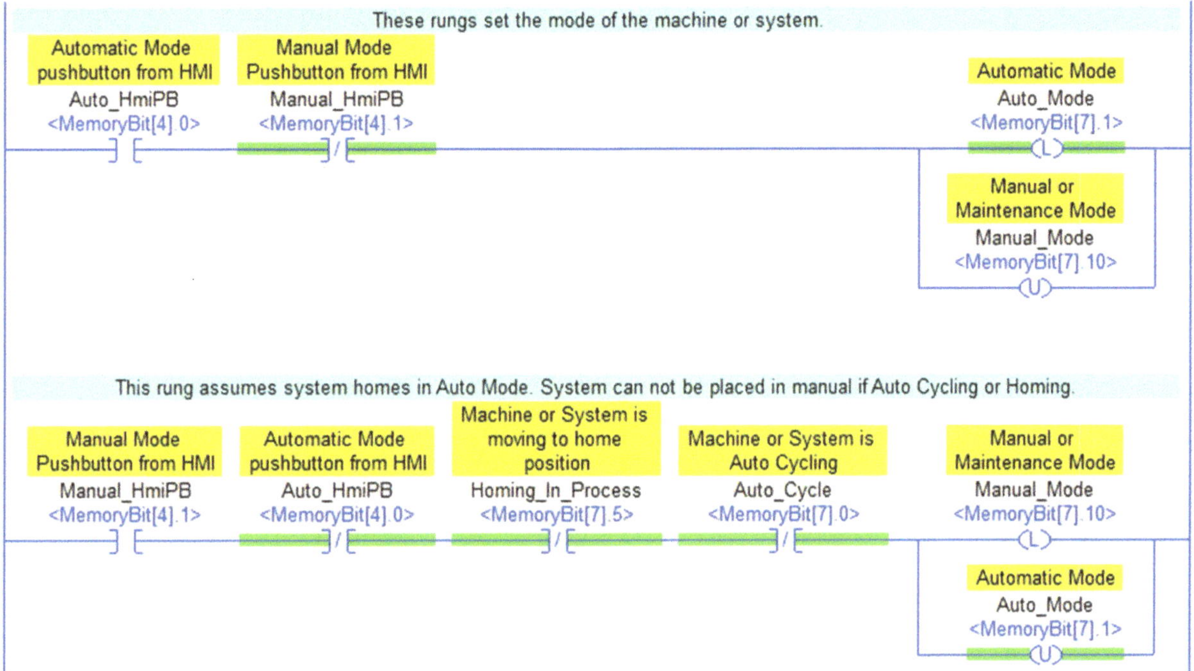

The diagram above shows some of the conditions for changing modes in a system routine. In this illustration, homing of actuators is done in Auto Mode, so the machine can't be placed in Manual Mode while in Auto Cycle or while homing.

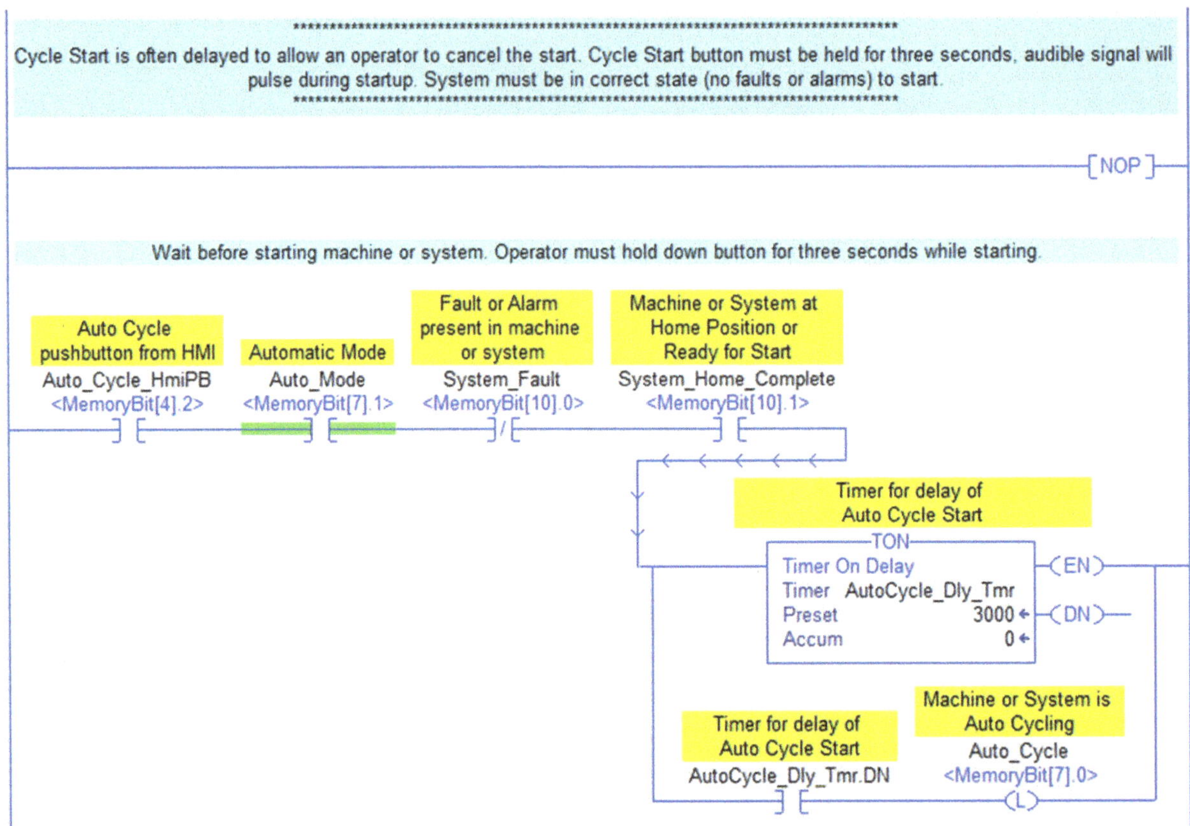

The diagram above shows a timer used to delay the start of a machine for three seconds. Usually, there will be an audible alarm that will sound while the operator presses the button, warning people in the area that the machine is about to start. This gives the operator a chance to remove his finger from the button and abort the startup if necessary.

Auto Cycle is used to allow a machine to automatically process components or material, it will usually do so until either a malfunction occurs or until an operator stops the process.

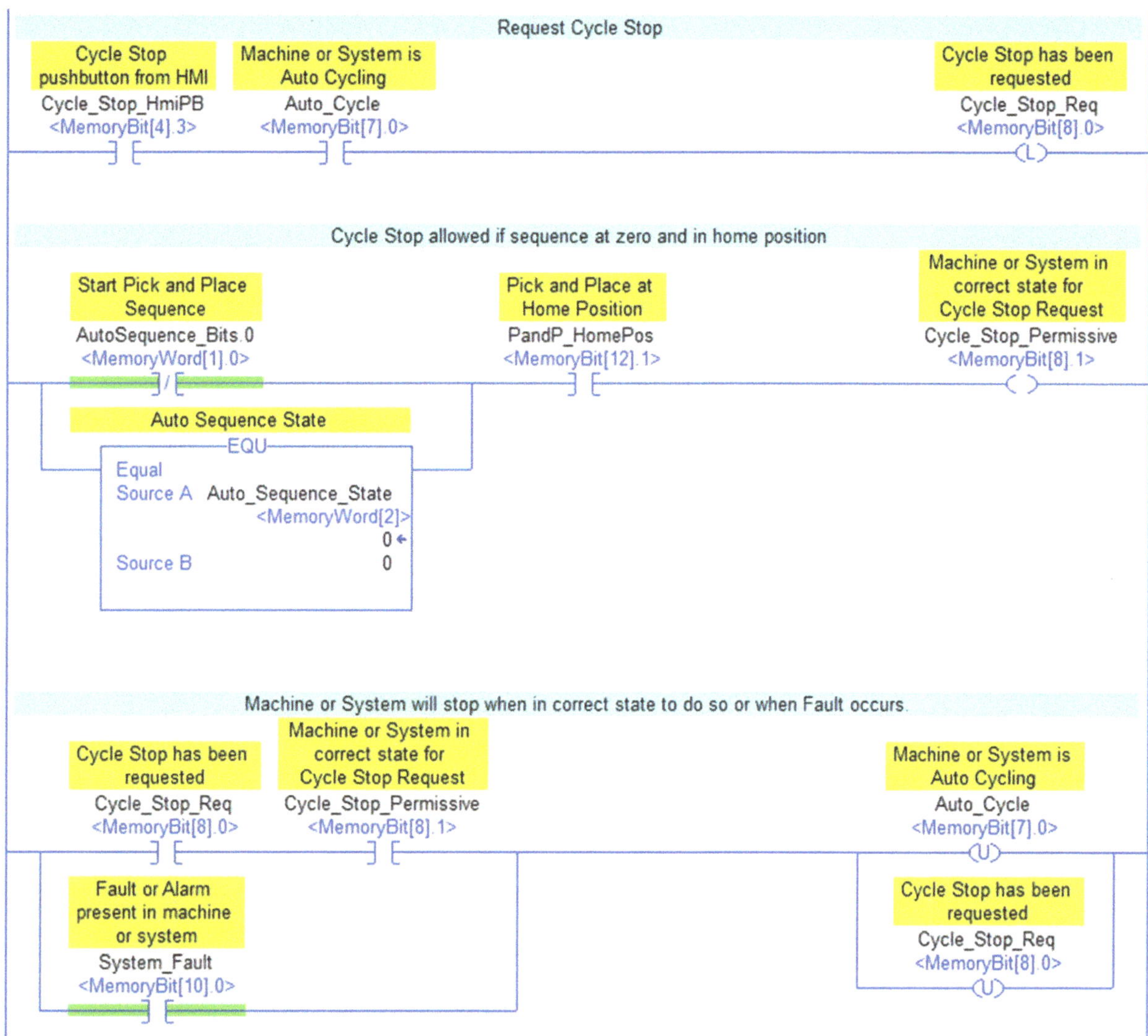

This shows logic for an operator to stop the system. If the system or machine is not in the correct state or position to stop, it will continue in Auto Cycle until the desired conditions or positions are satisfied; this requires that a "Request to Stop" bit is latched so that the system can remember the request. In this example, a fault in the system will stop the machine immediately.

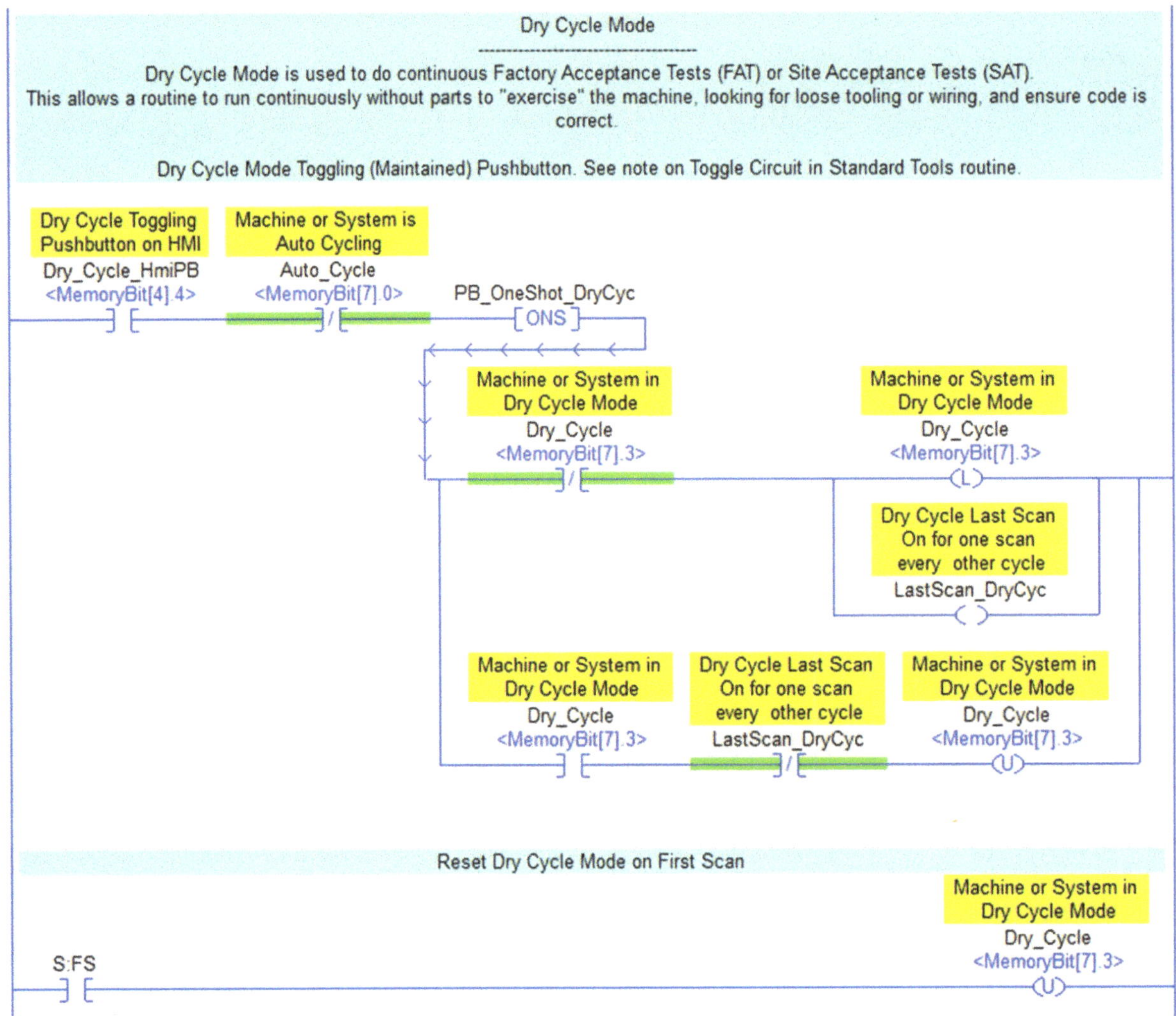

Dry Cycle is typically used in machine runoff conditions by machine builders, but the logic may be left in the program for the customer. The idea behind Dry Cycle is that the machine can run continuously without parts to "exercise" its actuators. The logic above incorporates a toggling pushbutton that will place the machine in Dry Cycle mode and take it back out with a single button. There is an additional example of a toggle in the "Tips and Tricks" section of this book.

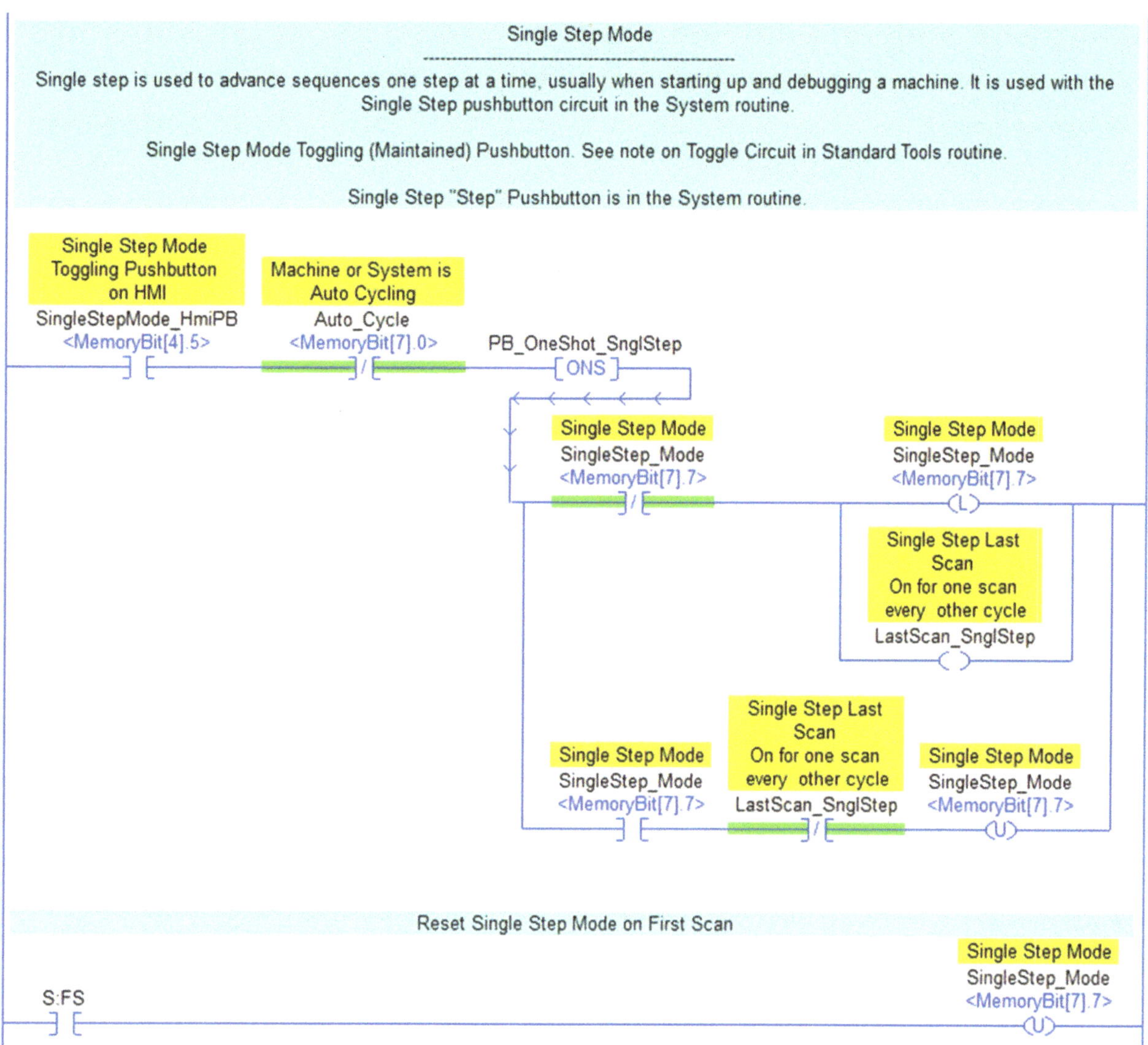

The logic above is similar to that of Dry Cycle mode control, it uses a single button to switch in and out of Single Step mode. Additional logic is used to create the bit that is placed in an auto sequence to step the sequence manually.

Note that both Dry Cycle and Single Step modes are unlatched when the PLC is powered off or goes into Program mode. This is optional.

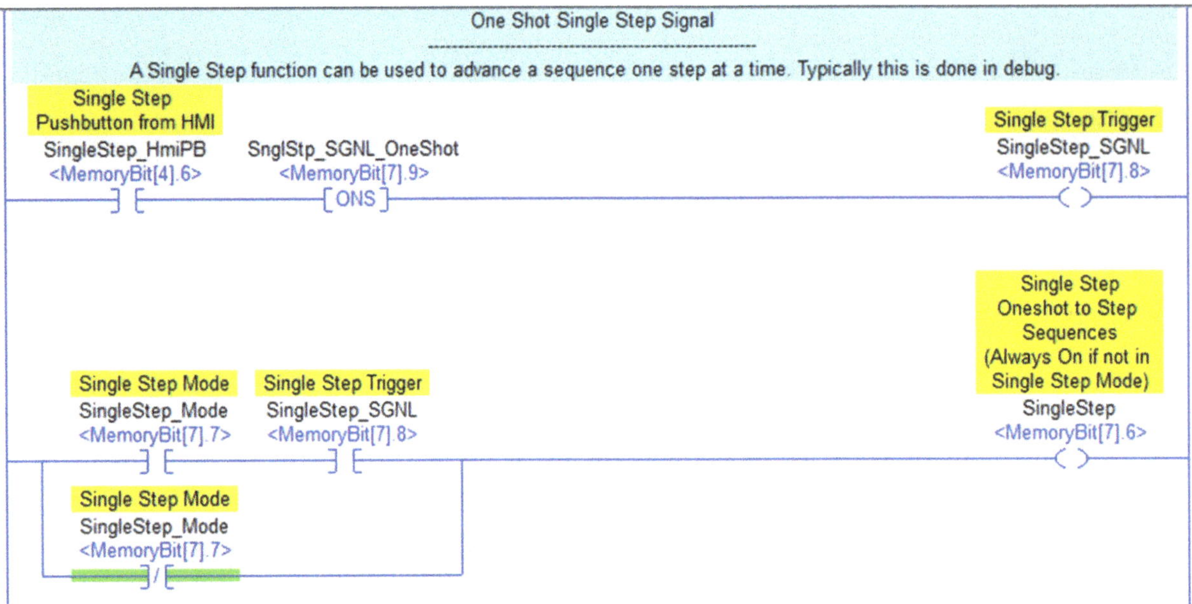

This shows how the signal that is used in the actual auto routine operates. When not in Single Step mode, the SingleStep bit is always on, and logic proceeds through the sequence as usual. When the mode is active, a pushbutton is used to move to the next step, provided the other conditions in the sequence step are satisfied. The Step signal is a one-shot.

For both Dry Cycle and Single Step modes, a complete understanding of how a sequence operates is necessary. Read through the **Auto Sequence Routine: Word Sequence** descriptions and then come back to this logic.

An **Input Routine** is used to modify the status of physical inputs for use in the program. Following are examples:

Back Checking. This is used for complementary pairs of sensors where they physically cannot be on at the same time.

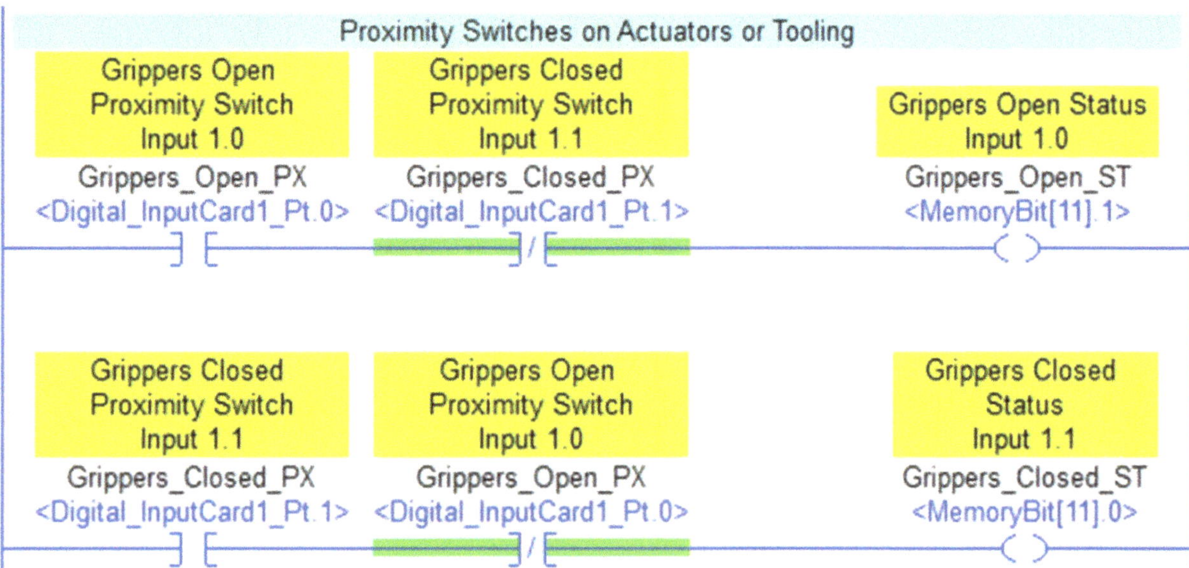

The ST or Status bits are used in the program rather than the physical inputs. This means the logic will not react if the incorrect signals are received due to a wiring or physical error.

Debouncing. This is used both to ignore intermittent, accidental signals and to place a delay on a signal to ensure that a part is settled into place before reacting.

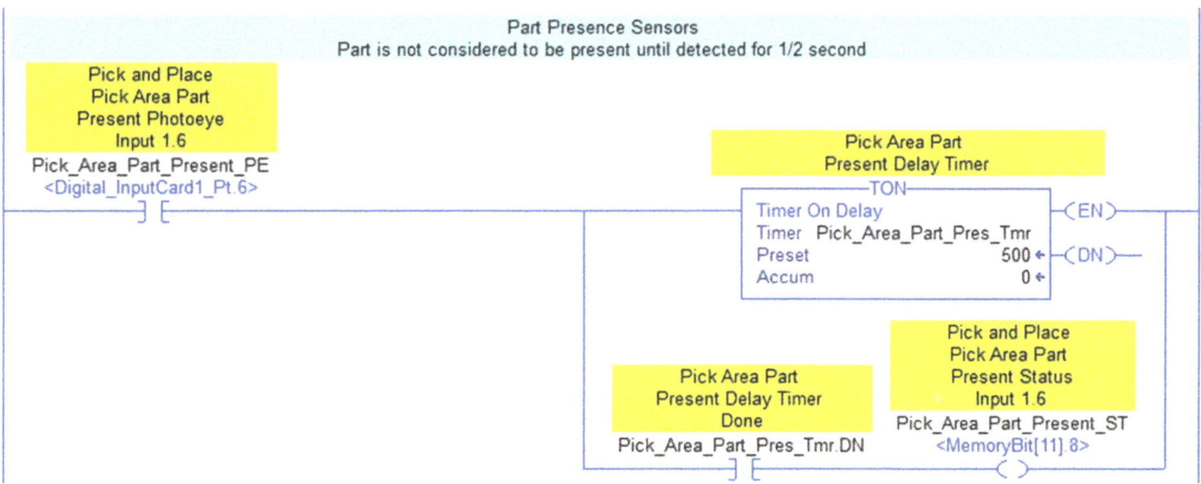

This is often used on sensors in conveyor systems where the part is conveyed against a stop before pushing it.

Listing the inputs and statuses in one routine also allows them to easily be located for monitoring or copying into other rungs of logic.

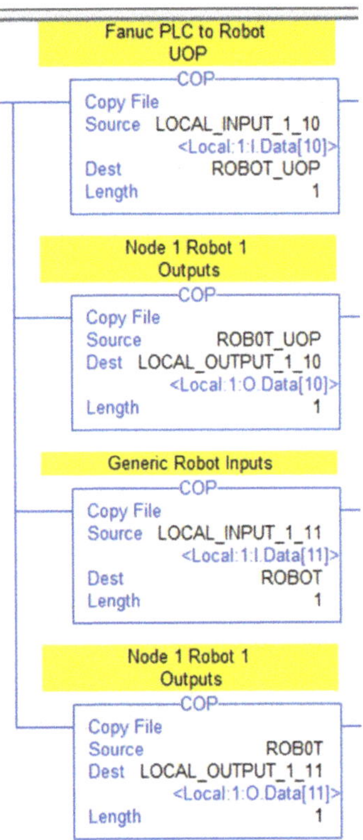

Mapping of I/O points is often done in order to interface a PLC with an external system. Files that represent complex systems, such as robots, variable frequency drives (VFDs) and machine vision are often imported as part of the hardware configuration. In order to add devices to the PLC hardware, an .eds (Electronic Data Sheet, Allen-Bradley) file is imported into the software. This is a text file, which among other things defines the input and output parameters of a device and their data types. While in some PLCs this automatically creates a tag with the I/O element of the device contained in it, sometimes a UDT needs to be created or the individual elements need to be mapped to PLC addresses.

Mapping is also common for interfaces between devices that communicate in different numerical bases, such as Siemens (byte-based) and Allen-Bradley (16 or 32 bit). Modbus communications also often require mapping.

Output Routines also usually contain mapping logic, as does an HMI Interface routine. HMI Indicator bits are often mapped when an "HMI Word" structure is used for individual indicators. HMIs may also have a list of internal tags that have to be mapped to PLC addresses.

Fault triggers are also often mapped into bits of words for HMI Alarms and Messages.

Output Routines: Because specific coils should only be placed in one location in a program, it is necessary to combine all of the different modes of operating the physical outputs in one location.

It is also important to ensure that outputs can't be energized if they can cause damage to equipment. Permissives are used to prevent outputs from activating if doing so is hazardous.

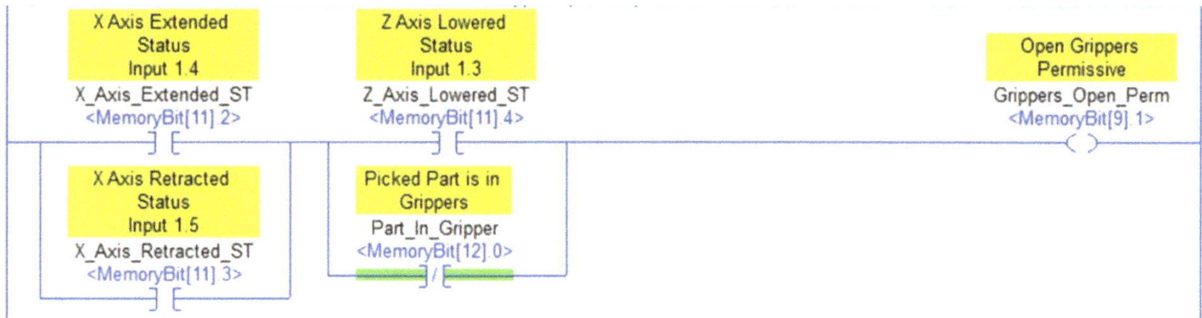

The permissive shown above prevents a gripper from opening unless it is at one end or the other of an X axis actuator. It also ensures that the part can't be dropped by inadvertently opening the gripper when it is raised. Permissives usually only apply in Manual Mode, since automatic functions have code preventing actions from happening in the wrong place.

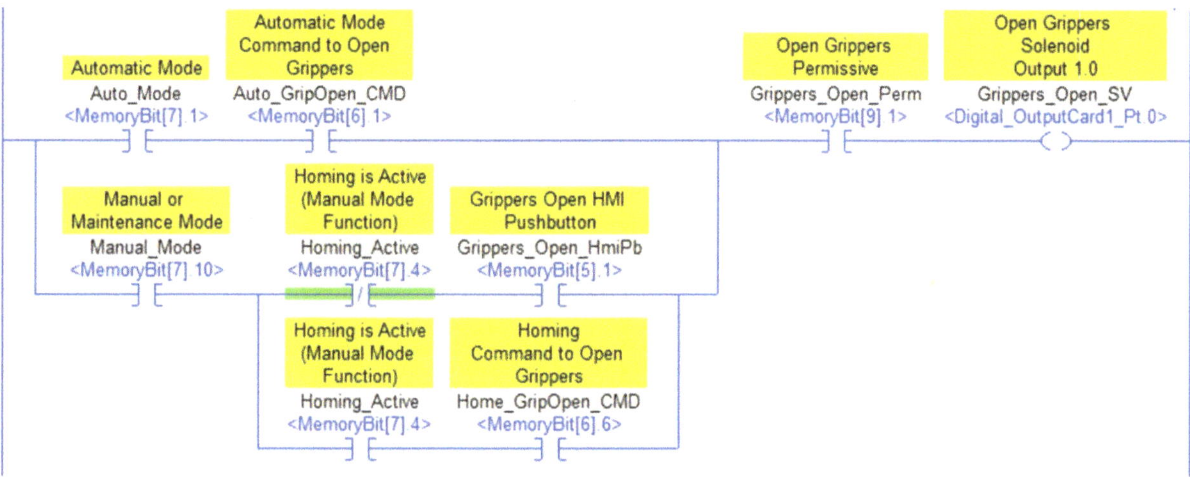

The output rung itself combines all the modes' commands into one location along with the permissive. In this case, homing is done in Manual Mode. The Auto Mode Command bit comes from the Auto Sequence, usually in its own routine.

As with the Input Routine, placing all the output rungs in one routine allows them to be easily located in the program. Both inputs and outputs should be listed in numerical order to make it easier for programmers and maintenance personnel to quickly find them.

Scaling functions for I/O should also be placed in the Input and Output Routines. Output mapping functions and sometimes HMI indicator mapping may also be placed here.

Faults and Alarms: As with the I/O routines, faults and alarms are often placed in their own routine so that they can be easily located. There are many different types of faults and alarms, and also different levels of severity. A fault may shut down the entire machine or system or disable only part it. Note that in the output structure on the previous page, the Auto Mode bit was used rather than the Auto Cycle bit. This means that if AutoCycle is reset, the actuator will still stay energized. The Auto Sequence will not progress however.

Alarms are also displayed on an HMI or SCADA screen. There are utilities in the HMI or SCADA software that allow for archiving of Faults (Alarm History), and acknowledging a fault, which silences alarms without clearing the fault condition. It is important to take this into account when creating the fault logic.

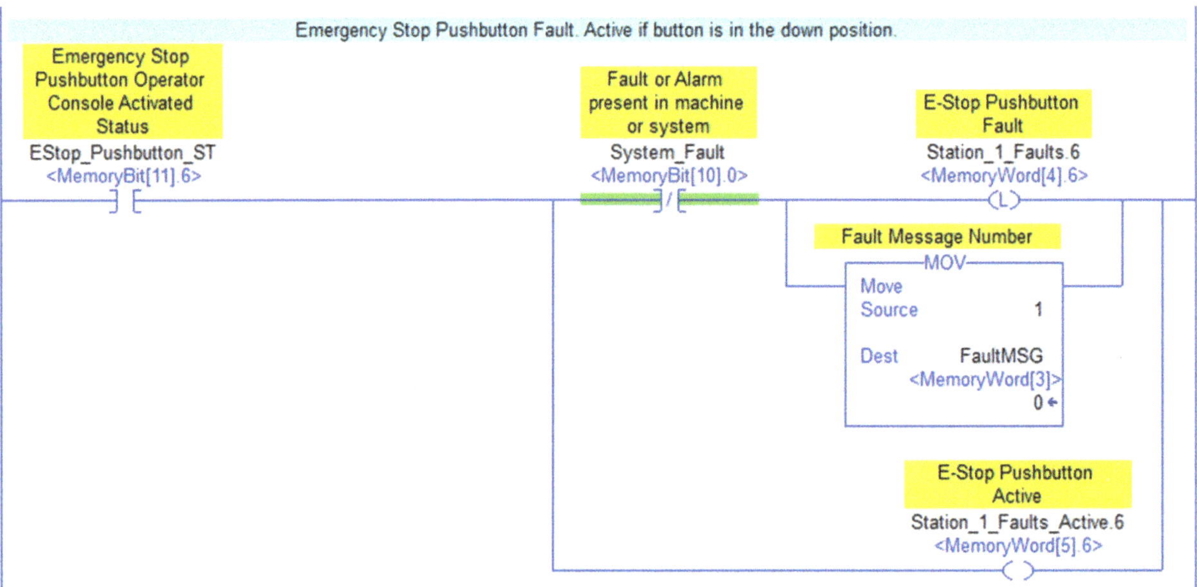

The fault logic shown above accomplishes several purposes.

1. It latches or sets a bit in the Station 1 Faults register. Even if the button is pulled back up, the fault stays latched.

2. It moves a number into the Fault message register; this is used to bring up a message in the HMI or SCADA system. The Station 1 Faults Active register is used to indicate whether the cause of the fault has been remedied. If any bit in the Station 1 Faults register is set, then the number will be non-zero; this is used to activate the System Fault bit, which is used in many different places in the program. Notice that in this logic, once the System Fault bit is true, it will prevent any new faults from being latched if it is placed into every fault rung.

It is important to ensure that the original cause of what might be a series of faults is retained. For instance, if an air cylinder jams on an operation, it will latch a fault. If someone opens a door or presses the Emergency Stop to address a safety issue, it will not create a new fault or move a message. If the air cylinder fault is reset after being corrected, a new fault will appear for the E-stop or door.

A common fault is to activate a timer when an output is on and stop it when the corresponding sensor is made.

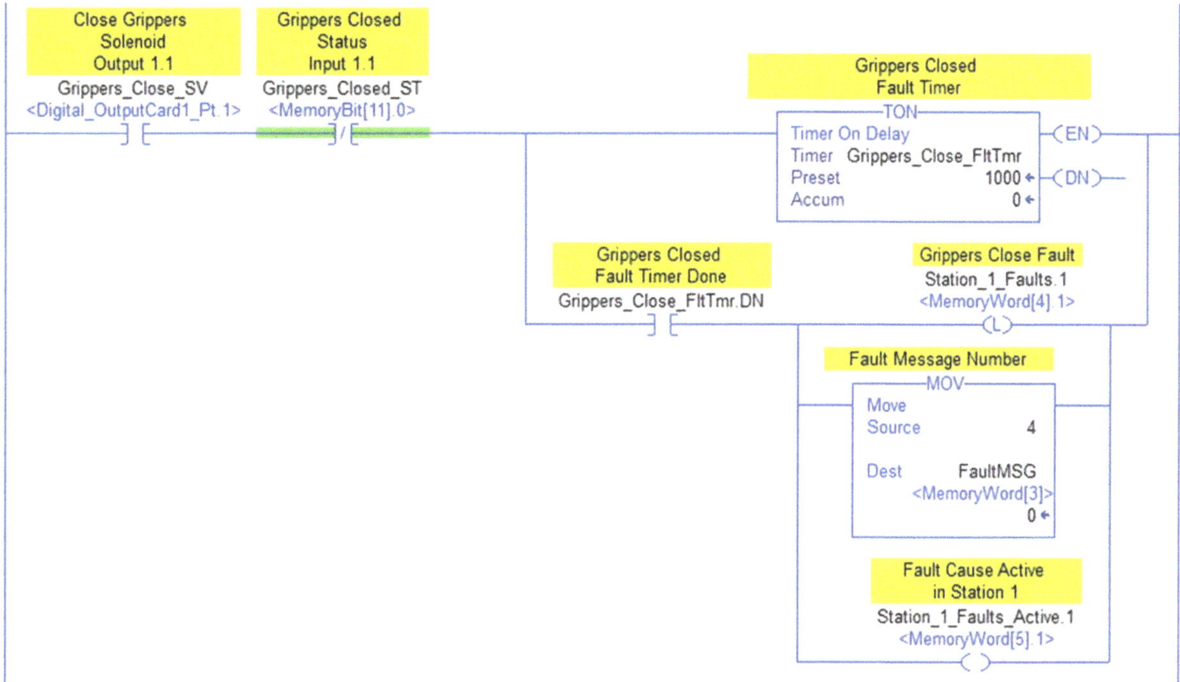

By using the back checked status bit mentioned in the Input Routine section instead of using the actual input address, both the closed and open sensors can be checked for faults.

If a normally closed System Fault bit is placed in series with the timer, as shown in the E-Stop Fault, the Active bit cannot be used to check the current status of the fault, but the timer will start again when the fault is cleared.

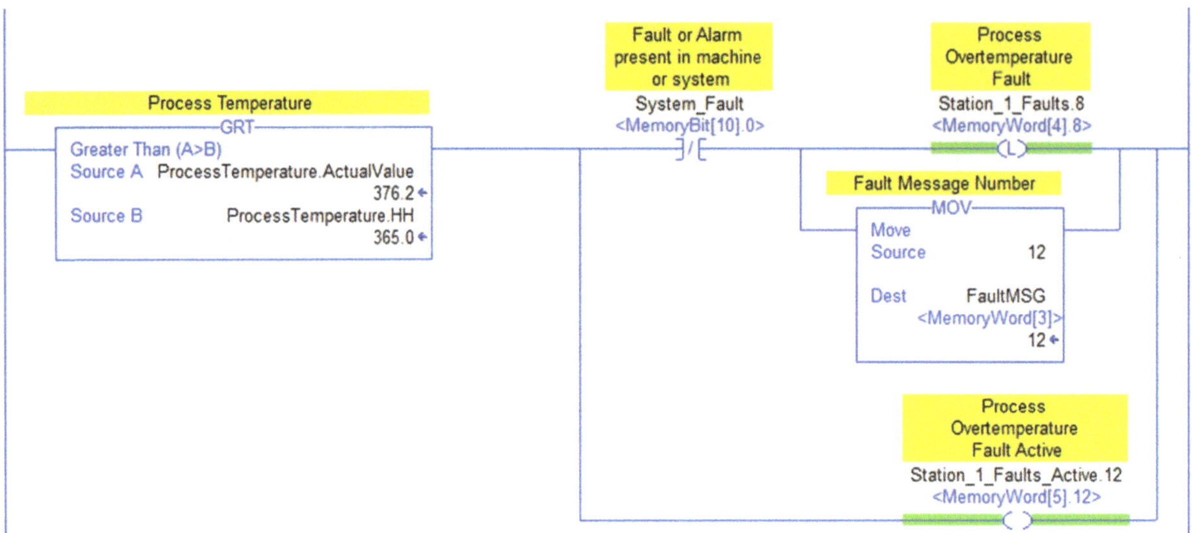

Process control faults often involve comparison logic that latches a bit if an analog input variable is above a limit. The fault again cannot be reset until the value has dropped below the alarm limit, as shown above.

Faults and alarms often have an audible alarm associated with them to alert an operator of the condition, as well as logic that prevents output devices from being energized and possibly damaging equipment. An operator can silence the alarm by acknowledging it on an HMI or SCADA screen.

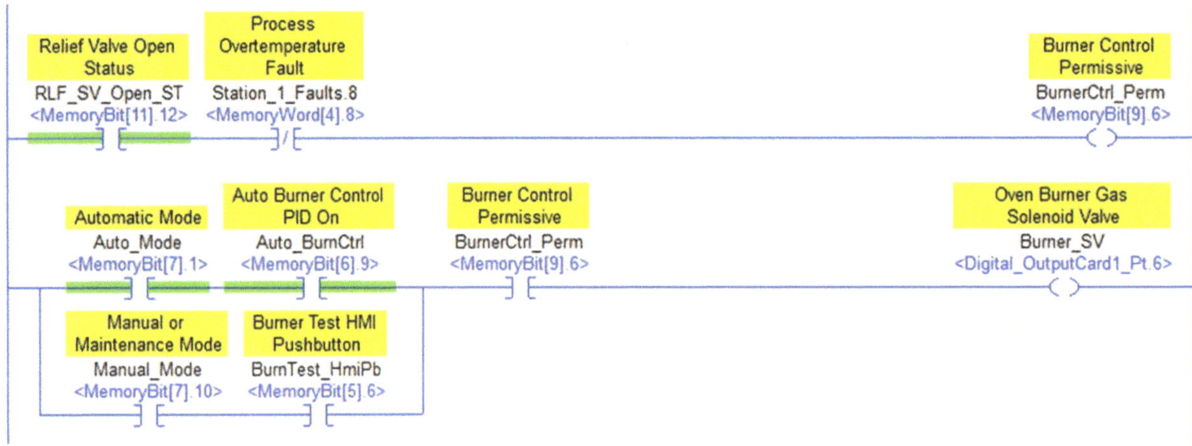

The logic above shows that the valve cannot provide gas to the burner if the fault is present. The Auto control bit is active, but the burner does not operate.

Faults are usually not placed in permissives or interlocks since hard-wired safety usually shuts off the actuator. This example illustrates the use of a specific fault disabling a specific device.

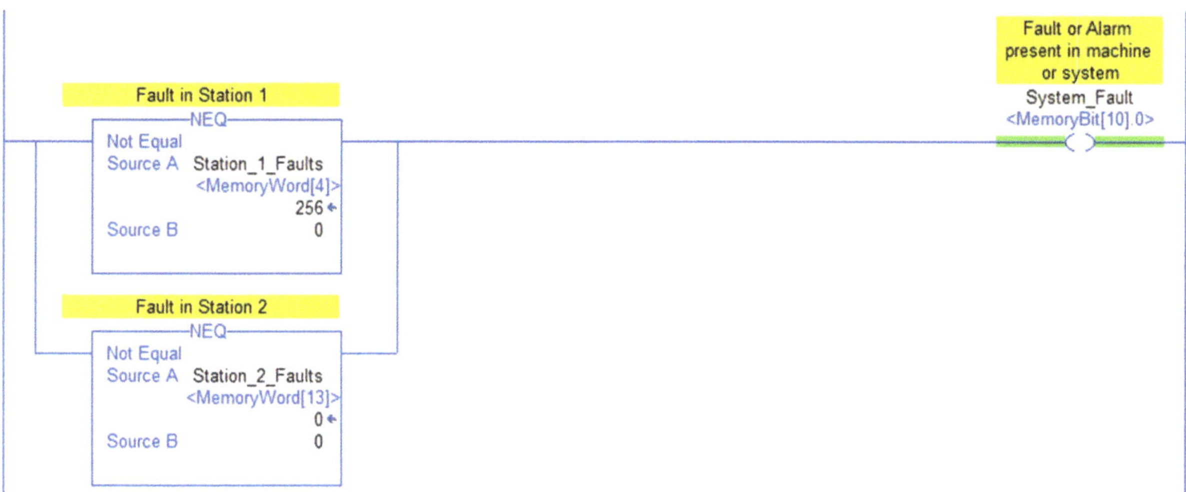

The overall status of faults and alarms can be summarized by comparing the bits as a group to zero as shown above. Faults are also often grouped by station so that if it is necessary for one part of a machine system to continue operating, it can be separated from another part.

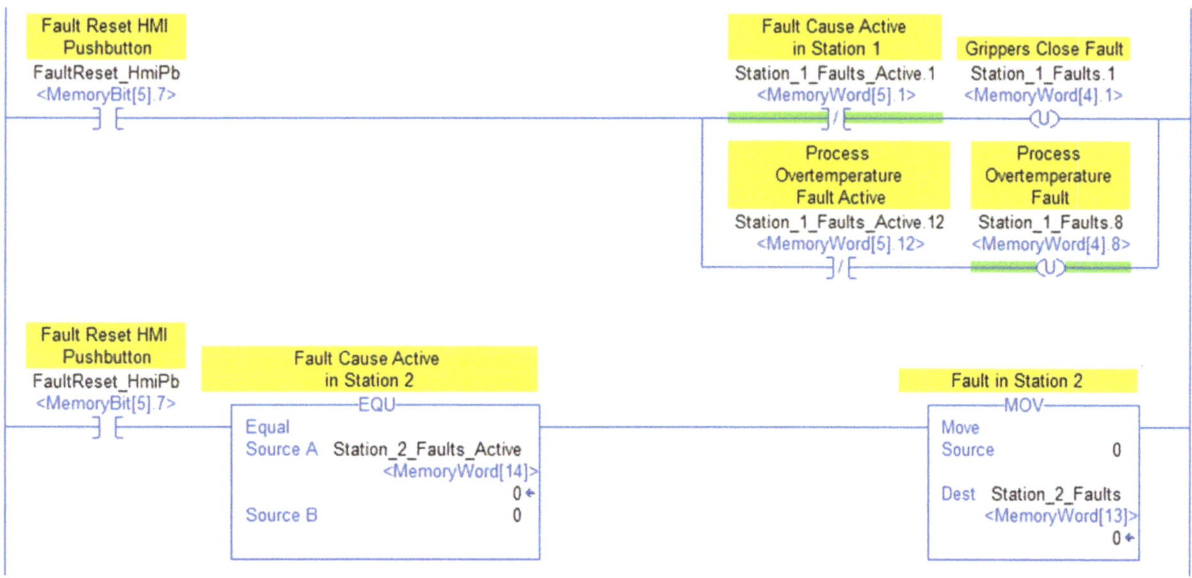

Faults and alarms can be reset individually based on their active conditions, or they can be reset as a group based on the station's condition. The programmer must do a detailed analysis of the effects of resetting faults and alarms.

A zero can also be moved into a Fault Message register when the fault is reset. The message register would then display a message such as "No Faults In System" on an HMI. If there are multiple faults in a system or machine, scrolling message displays or an alarm history screen can be used.

The **Auto Sequence** is the heart of an automated control system. It requires more planning and analysis than the other types of routines, and there are several methods of programming them.

The first method is often used by beginner programmers, as it is easy to understand and implement.

In the simple bit sequence on the next page, a bit is set or latched based on a sensor or other condition. While the bit is on, another bit, the Auto Output bit, is used to energize an output while in Autocycle or Auto Mode. Once another input or status condition is active, often due to the output being energized in the previous step or event, the step is unlatched or reset and the next step is latched on.

The Auto Output bits in this example are used in the position of the CMD or command bits in the section on output routines.

Programmers often latch and unlatch bits to control outputs in Auto Mode without realizing that they are actually writing a sequence. This can become very confusing if a list of events or flow chart is not used to plan the sequence ahead of time. Note that in this example, each output control is energized during only one step; this can easily be modified by adding conditions.

Bit Sequence: The previous example is a simplified version of what is sometimes called a "bit sequence". If the bits of a word or double word are used, the sequence can be easier to control and follow.

In the example above, a pneumatic actuator is used to push a part off of a conveyor when detected by a photoeye. Notice that the sequence bits are all in the same memory word, making it possible to unlatch all of the bits with the MOVE command at the end. The previous bits are not unlatched at each step as in Figure 25; instead, they are set or latched sequentially and then cleared at the end of the sequence.

Using the *Auto_Cycle* bit in every rung ensures that the sequence will not proceed if a fault unlatches the bit. By placing the normally closed contact of the next step bit the sequence will not evaluate the logic of previous steps. This makes it operate similarly to the Word Sequence described next.

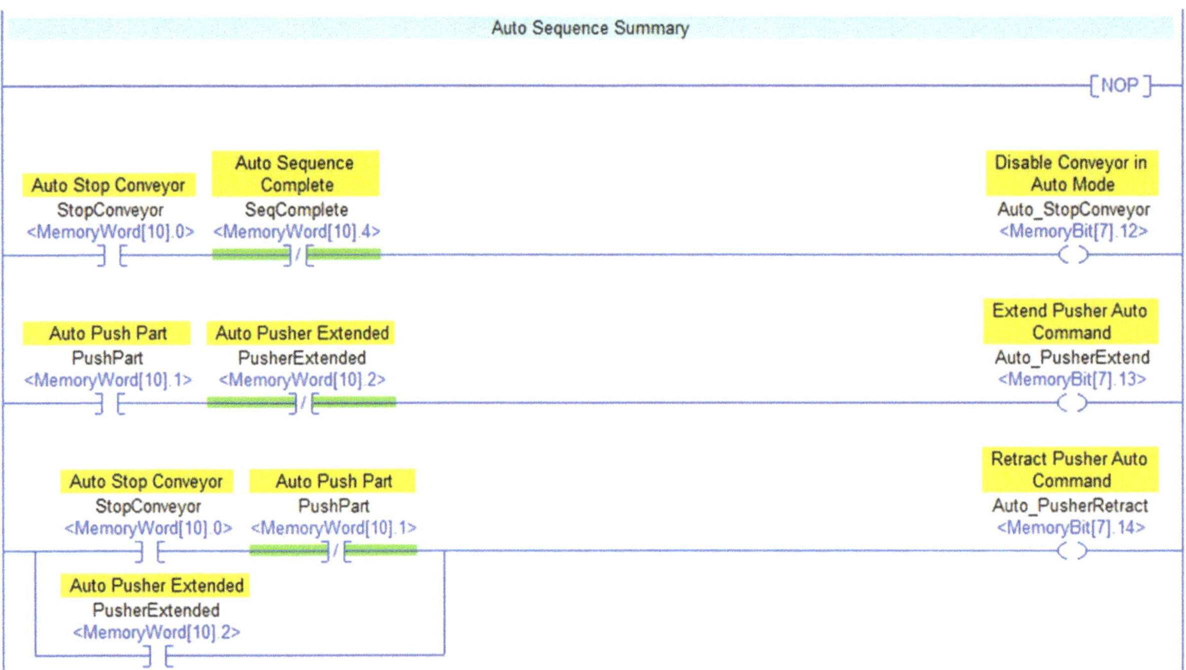

The control bits used in the Output Routine are often **summarized** after the sequence itself. This can make troubleshooting the sequence easier and improves "readability" by not separating the steps.

Bit sequences can still be somewhat confusing because it is possible to be in more than one step or state at a time. This can be both an advantage and a disadvantage; it is possible to reset bits in part of the sequence while leaving other bits latched, creating parallel sequences. It is important to fully document this type of sequence completely, explaining what the steps are doing.

Word Sequence: Another common method of writing sequences is to move values into an integer register.

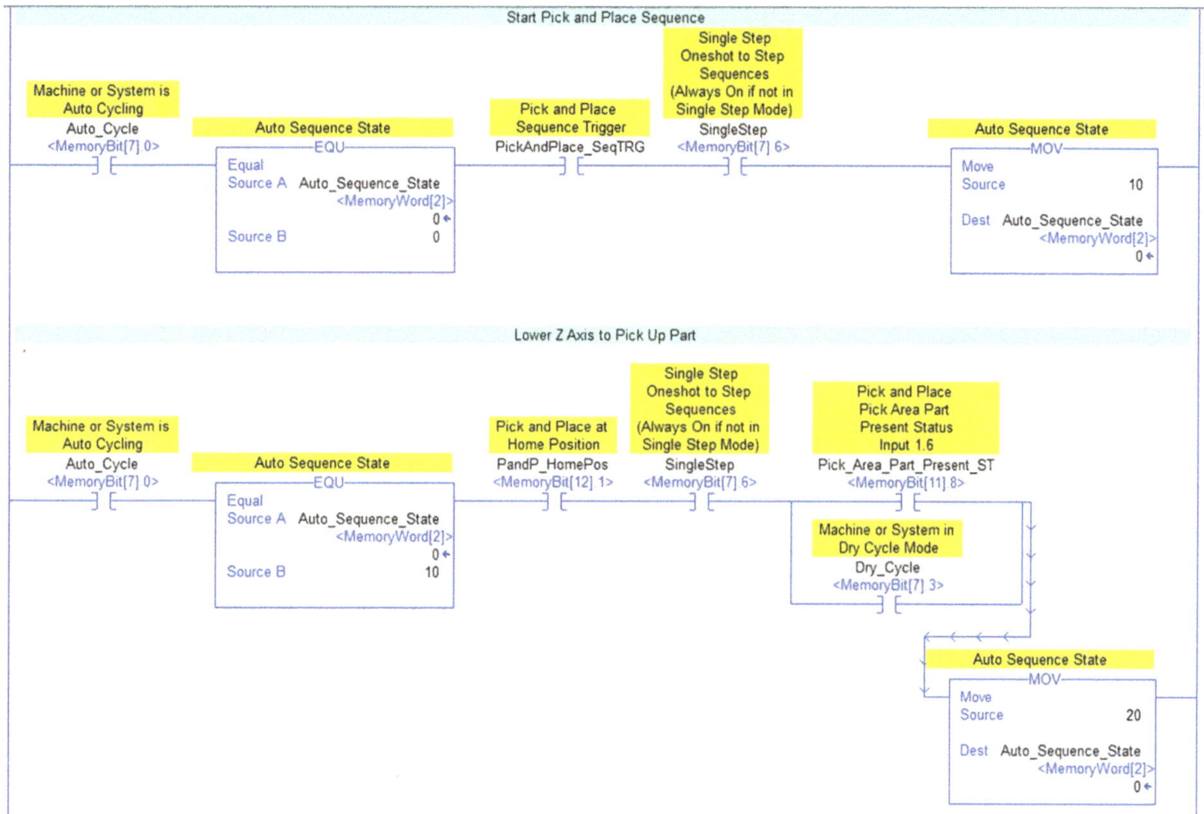

The Pick and Place logic on the next couple of pages shows how the sequence progresses. Rather than latching bits, numbers are moved into a register defining the current state.

This sequence also shows the use of Single Step and Dry Cycle modes. Single Step allows an operator to step the sequence one stage at a time, while Dry Cycle lets the sequence run without parts being present.

Once again, if the machine faults, the system will drop out of AutoCycle and will not advance. It will maintain its current state unless the fault is cleared and placed back into AutoCycle or it is homed.

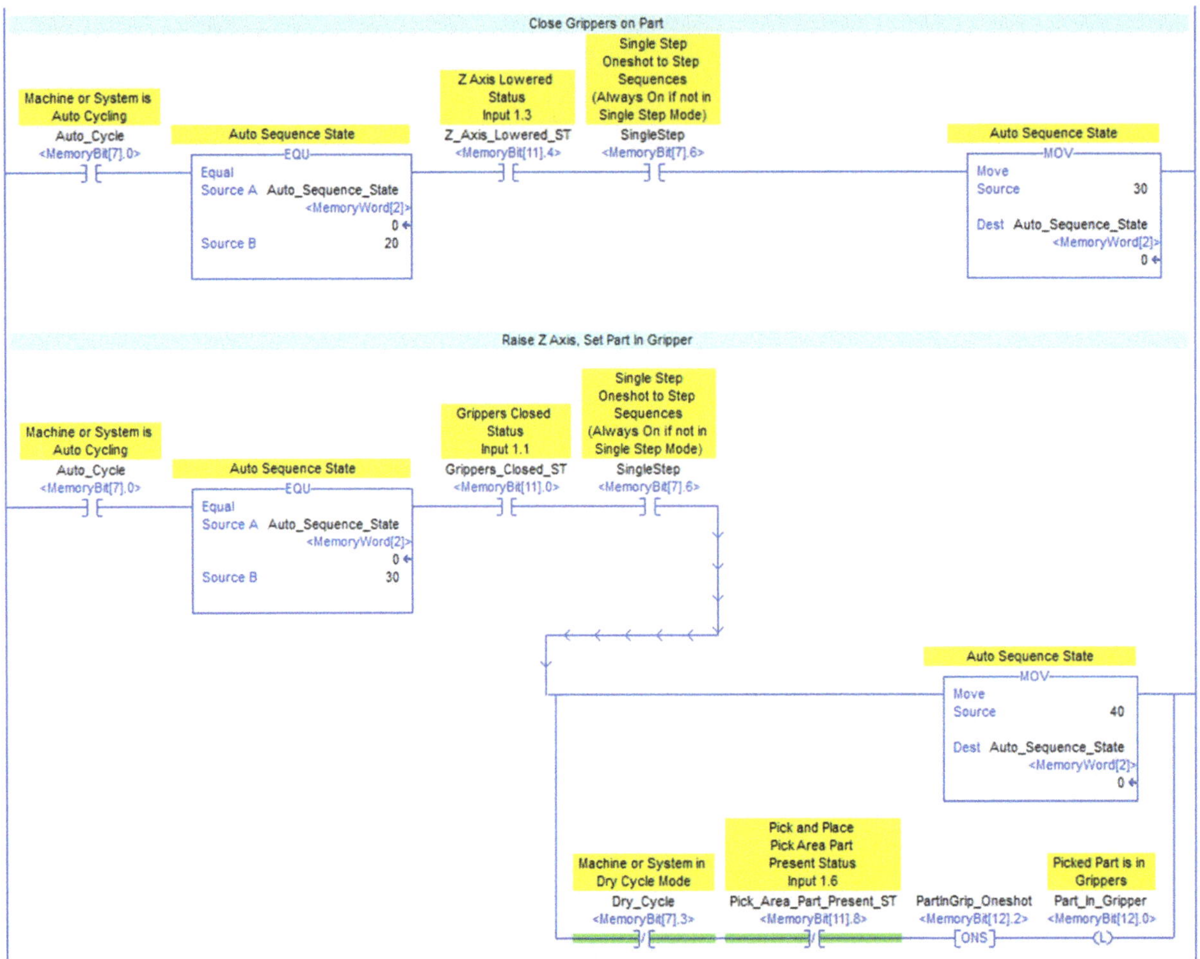

Note that during Step 30, a memory bit is set, verifying that the part was removed and is now held in the grippers. This can be important when determining whether it is permissible to open the grippers during a homing routine or in part tracking.

Gripper sensors are also sometimes positioned such that if a part is present, the grippers don't close all the way and the sensor stays on. If the part is not present, the grippers pass the sensor; because of this, a "debounce" timer is often used for gripper sensors. With a missing part, the sensor will not be made long enough for the sensor to expire.

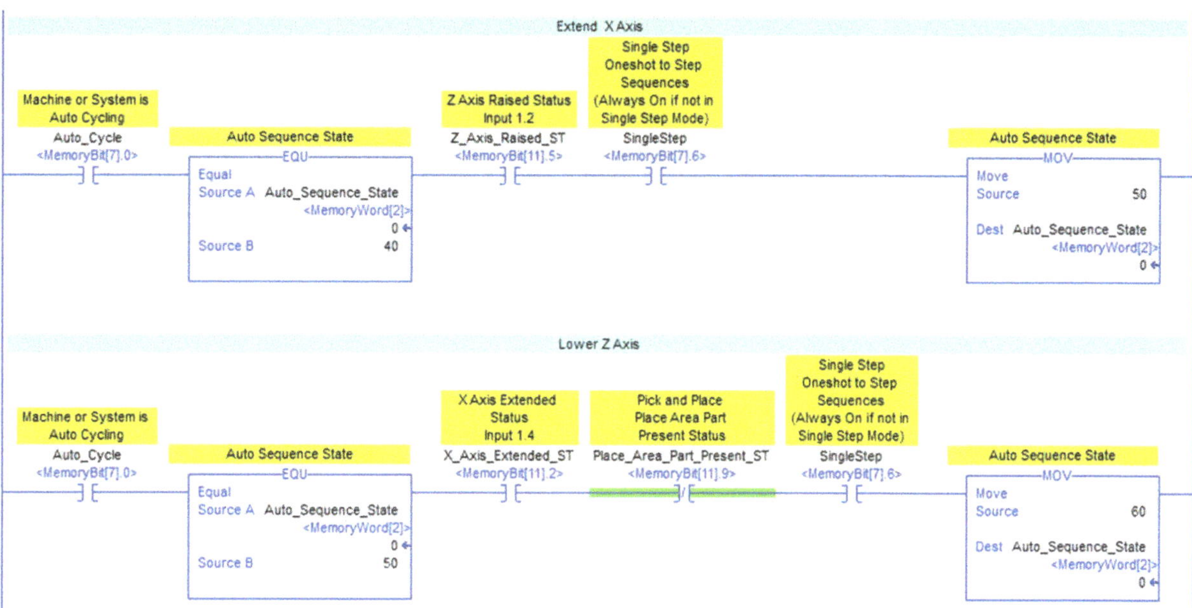

Step 50 ensures that a part is not already present in the location where the part is to be placed. This acts as a permissive for setting the part down. If there is already a part there, the sequence will wait to advance to Step 60 until it is removed. This is done by an operation outside of this sequence; multiple sequences are often written to operate machine actuators that may have to interface with each other.

It is unnecessary to use the Dry Cycle mode bit to bypass the Part Present sensor, since the part should not be there.

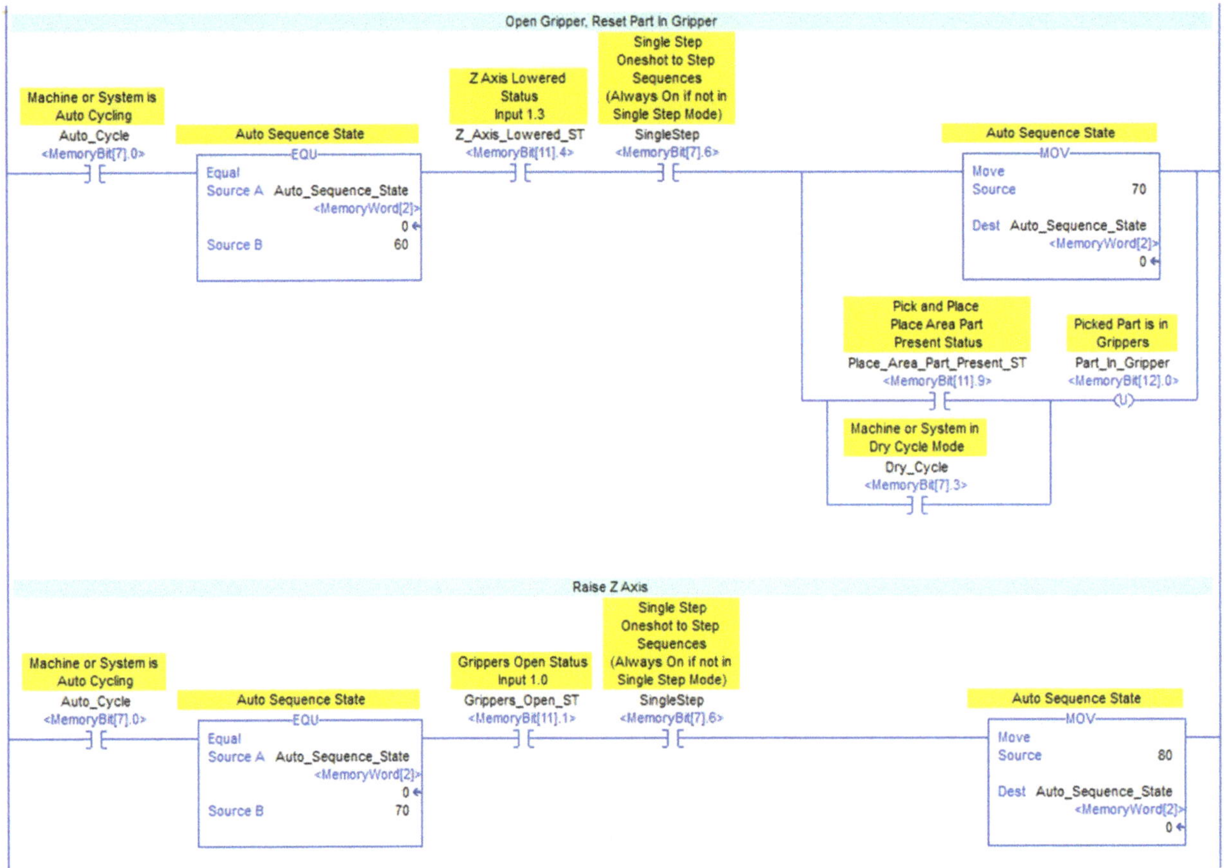

As with Step 30, a memory bit is used to determine whether a part is still in the grippers. The bit is unlatched in Step 60 once the part is set down.

In Dry Cycle Mode, the Part Present sensor is bypassed and ignored. The "Part In Grippers" status memory bit will automatically be set.

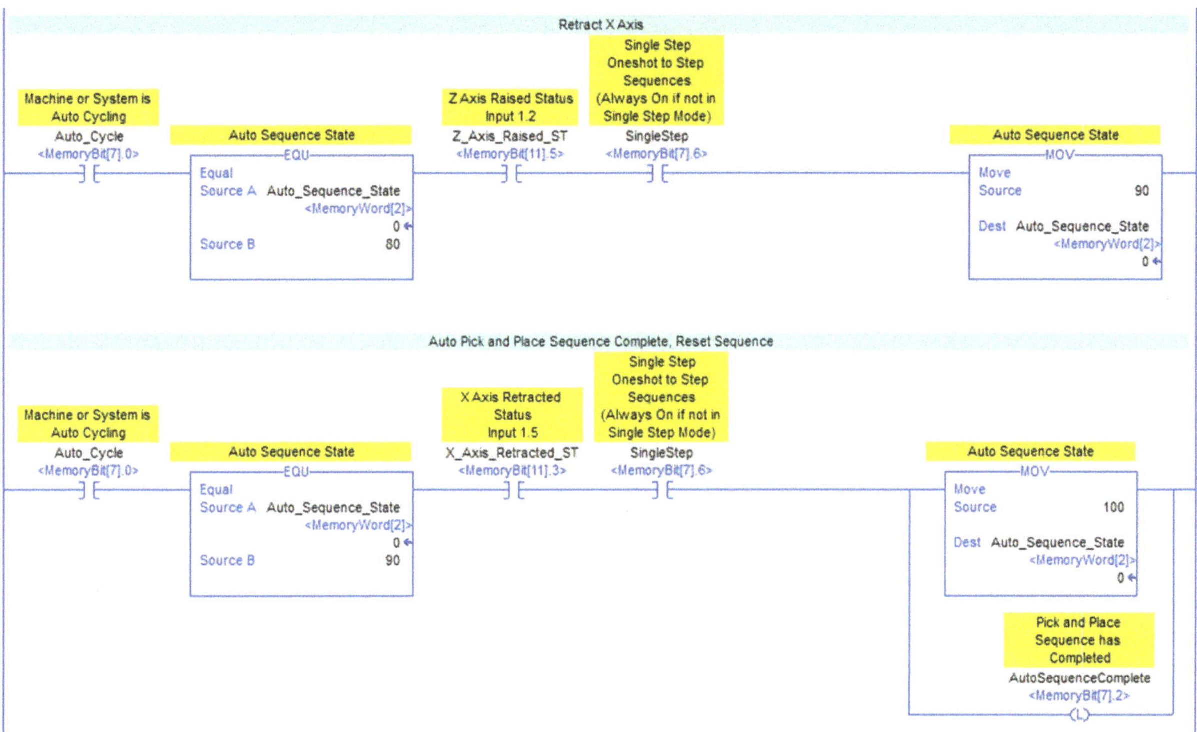

The sequence ends with the Pick and Place mechanism back at the home position. A Sequence Complete bit is latched and can serve as a trigger for other operations or sequences.

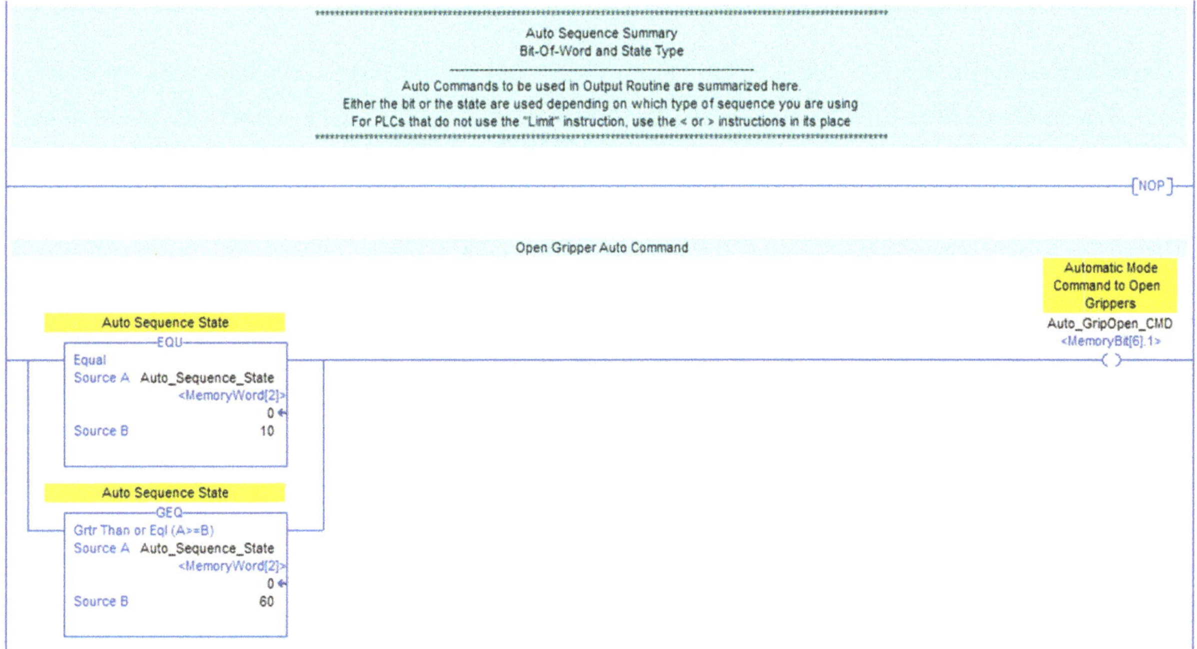

As with the bit sequence, summary bits are often listed after the sequence and are then used to drive outputs in the Output Routine.

Word sequences are different than bit sequences in several ways. In a word sequence, there is only one operational state active at any time. By sequencing in steps of ten or even one hundred, spare states can be built into the sequence.

In bit sequences, the states or steps can be individually named, but a machine can have any number of bits latched or unlatched at any given time. This can make them more flexible, but harder to follow the logic and debug.

Similar to a word sequence is a **State Machine**. Modes and sequence steps are sometimes combined to describe not only the step of the operation, but also conditions such as Starting, Started, Aborting, Aborted, Stopping, Stopped and Faulted. Ranges of numbers may be assigned to these states such as 5000-6000 for Auto Cycle states, 9000 for Faults, and <1000 for startup and initialization. There are several templates used by large manufacturers that use this technique. It is also a common method for robot control.

Homing Routines

Homing is used to ensure that actuators and machine conditions are brought to a known and appropriate starting position. Often when machinery is in AutoCycle and a fault condition occurs, operators may need to move actuators manually or remove in-process parts. This means that the Auto Sequence is no longer in the correct condition to continue. In this case there are several things that must be considered:

1. The out of position or missing part condition must be detected. For actuators, this can be done by determining the correct bit pattern for a given sequence step or state and comparing it with the actual input state, as shown below. Another method is to check whether the machine was placed into manual mode or if manual operation using pushbuttons occurred while the Auto Sequence was active.

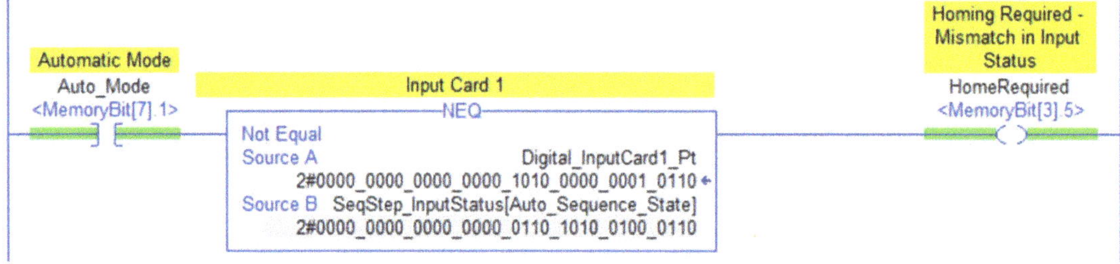

2. The Auto Sequence affected must be aborted or brought back to the correct state, usually zero.

3. The programmer needs to determine whether the homing operation will occur in Auto Mode, Manual Mode or both, or if an entirely new Homing Mode needs to be defined.

4. A Homing Sequence needs to be written. This uses the same technique previously described in the Auto Sequence section. Actuators need to be moved in order, returning them to their appropriate positions so as not to damage equipment. For example, in the previous Pick and Place sequence, the Z axis must be raised before the X axis is returned. If there is a part in the grippers, a decision must be made as to whether the operator must remove the part manually, or if the part can safely be placed in either the pick location or the place location.

Homing Sequences are every bit as complex and necessary as Auto Sequences are. Down time in machinery is often traced to operators not being able to diagnose why the system can't continue, which is also related to insufficient diagnostic messaging.

Tips, Tricks and Techniques

Recipes

A recipe is not necessarily a list of ingredients in a process like most people are familiar with in the kitchen. It also contains machine states, timer setpoints, bits that determine whether a part needs to be worked on in a station, and, of course, process values.

Recipes	udt_Recipe[8]	Recipes
+ Recipes[0]	udt_Recipe	Recipes Recipe File for Part
− Recipes[1]	udt_Recipe	Recipes Recipe File for Part
+ Recipes[1].Name	STRING	Recipe File for Part Recipe Name
+ Recipes[1].C1_PartNbr	DINT	Recipe File for Part Assembly Part Number
+ Recipes[1].Component	SINT	Recipe File for Part Component Bits
+ Recipes[1].Weight	REAL[8]	Recipe File for Part Component Weights
+ Recipes[1].C2_PartNbr	DINT	Recipe File for Part Component 2 Part Number
Recipes[1].C3_Torque	REAL	Recipe File for Part Component 3 Torque
+ Recipes[1].C6_Color	INT	Recipe File for Part Component 6 Color
+ Recipes[2]	udt_Recipe	Recipes Recipe File for Part
+ Recipes[3]	udt_Recipe	Recipes Recipe File for Part
+ Recipes[4]	udt_Recipe	Recipes Recipe File for Part
+ Recipes[5]	udt_Recipe	Recipes Recipe File for Part

The figure above shows an array of 8 recipes using a UDT (User Defined Data Type) that describes a part. This part is also used in the part tracking routine.

The recipe information includes the name of the recipe itself, which in this case is also the part number for the assembled part in STRING format. It also includes decimal part numbers for sub-components in DINT form, a target torque value as a REAL number, and an array of REAL numbers corresponding to the approximate target weight at several assembly stations in the production line. A part color is represented by an integer, and a byte or SINT is used to track component presence in bit form.

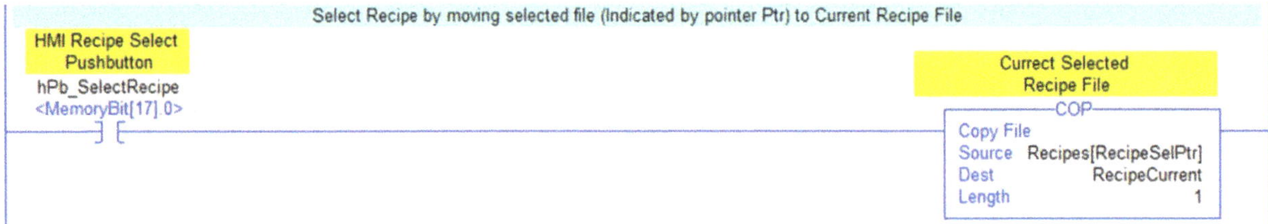

The logic above uses a pointer called *RecipeSelPtr* to indicate which Recipe file is moved into *RecipeCurrent*, which is then used in production logic in other routines. This is an example of **indirect addressing**. When the HMI button is pressed, the selected recipe is moved. The selection is an integer numeric entry field on the HMI screen that identifies the recipe by number. A corresponding ASCII field on the HMI displays the *Recipe[ ].Name* field to verify the recipe selection.

If the recipe needs to be changed or edited from the HMI, the logic below can be used.

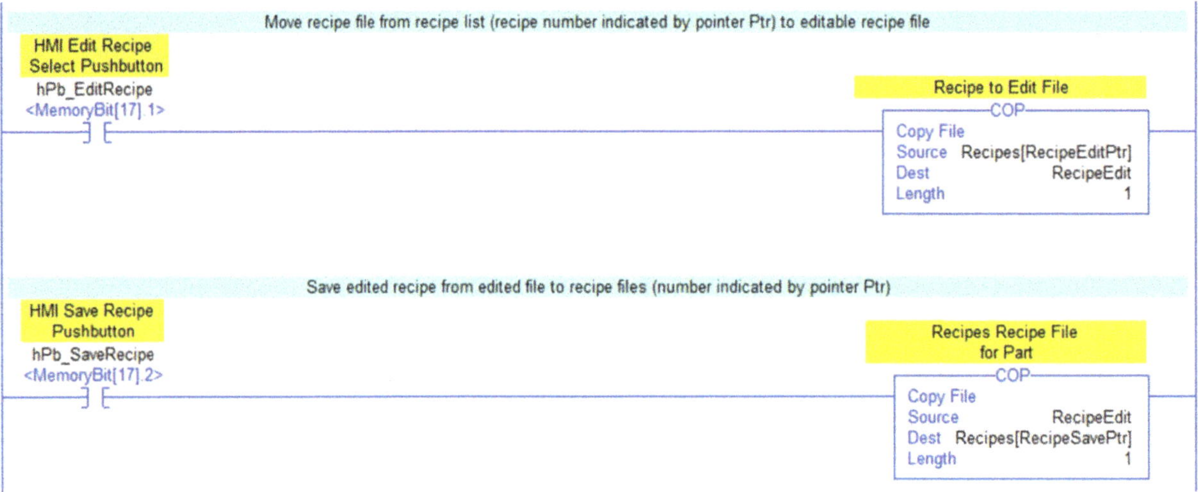

The *RecipeEdit* tag is another instance of the *Recipe* UDT. Entry fields on the HMI correspond to each of the items in the recipe. After changes have been made, the Save Recipe pushbutton moves the edited recipe into the recipe number designated by the Recipe Save pointer (*RecipeSavePtr*)

In a ControlLogix processor, entering a number outside of the limits of the array dimensions into the pointer will fault the processor. The logic below prevents this from happening.

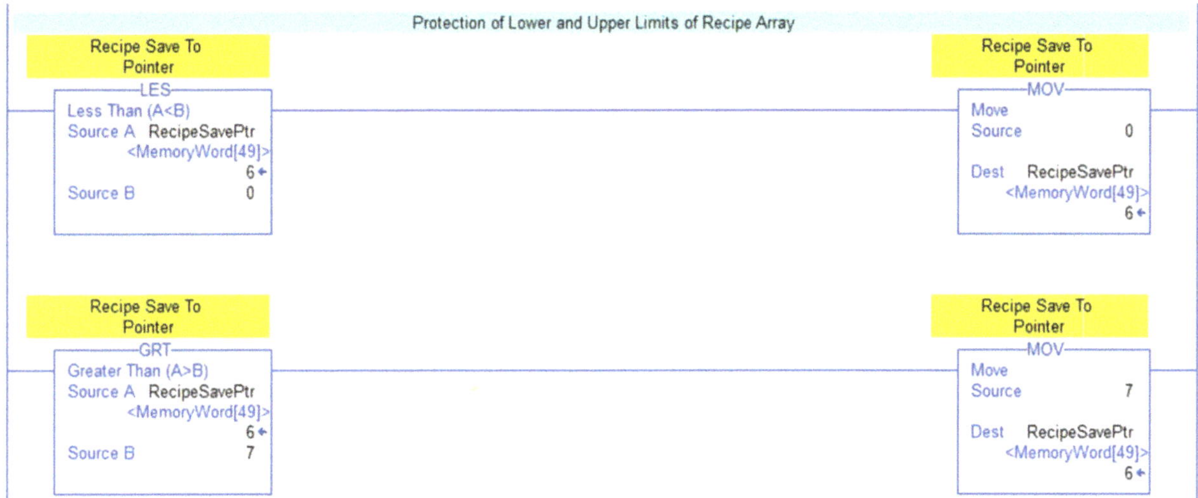

Many HMI and SCADA platforms have recipe management utilities built into their software. Recipe parameters and files also are often comma delimited .csv files that can be edited with external packages such as Microsoft Excel. This simple method of recipe management works well for simpler HMIs. For PLCs that don't have UDT or Array capability, moves need to be made on an individual register basis.

Part Tracking

For machines that process discrete parts in assembly operations, it is important to ensure that data associated with the part is carefully documented and that the data follows the part through the system. It is also not always possible to use bar codes or other identification systems to read the part at every station. Because of this, various techniques can be used to move data as the part proceeds from station to station.

StationPart	udt_Part[10]	Part by Station Array
+ StationPart[0]	udt_Part	Load Station 1
+ StationPart[1]	udt_Part	Assembly Station 2
− StationPart[2]	udt_Part	Torque Station 3
StationPart[2].Weight	REAL	Torque Station 3 Weight of Component
+ StationPart[2].Component	SINT	Torque Station 3 Component Installed Bits
+ StationPart[2].PartNbr	DINT	Torque Station 3 Part Number
+ StationPart[2].PartCode	STRING	Torque Station 3 Serial Number from Bar Code
+ StationPart[2].FailCode	SINT	Torque Station 3 Failure Code Byte
+ StationPart[2].Station	DINT	Torque Station 3 Station Complete
StationPart[2].OK	BOOL	Torque Station 3 Part OK Inspection Pass
StationPart[2].NG	BOOL	Torque Station 3 Part NG Inspection Fail
StationPart[2].InProc	BOOL	Torque Station 3 Part In Process
StationPart[2].Complete	BOOL	Torque Station 3 Part Complete
+ StationPart[3]	udt_Part	Torque Test Station 4
+ StationPart[4]	udt_Part	Reject Station 5
+ StationPart[5]	udt_Part	Assembly Station 6
+ StationPart[6]	udt_Part	Gasket Station 7
+ StationPart[7]	udt_Part	Inspect Station 8
+ StationPart[8]	udt_Part	Reject Station 9
+ StationPart[9]	udt_Part	Unload Station 10

The information in the array not only needs to be moved from station to station as the part moves through the assembly process, it also needs to be modified at each station. The figure above shows the structure of the part data. The Weight variable is updated at each station as the pallet is weighed; this information is used to help ensure that the proper components have been added. As components are added, the Component bits are set at the corresponding station. The *PartCode* variable is obtained from a bar code reader at the first station, which corresponds with the part number in the recipe. The *OK* and *NG* bits are updated at each inspection station; if the NG bit is set, then the part will be rejected at Stations 5 and 9. The cause of the failure can be encoded into the *FailCode* byte (SINT).

The InProc and Complete elements of the UDT are used to determine when the movement of parts will occur. For this example, a simple 10-station assembly line is used to illustrate some of the concepts in tracking parts. Station names and functions are listed as descriptions.

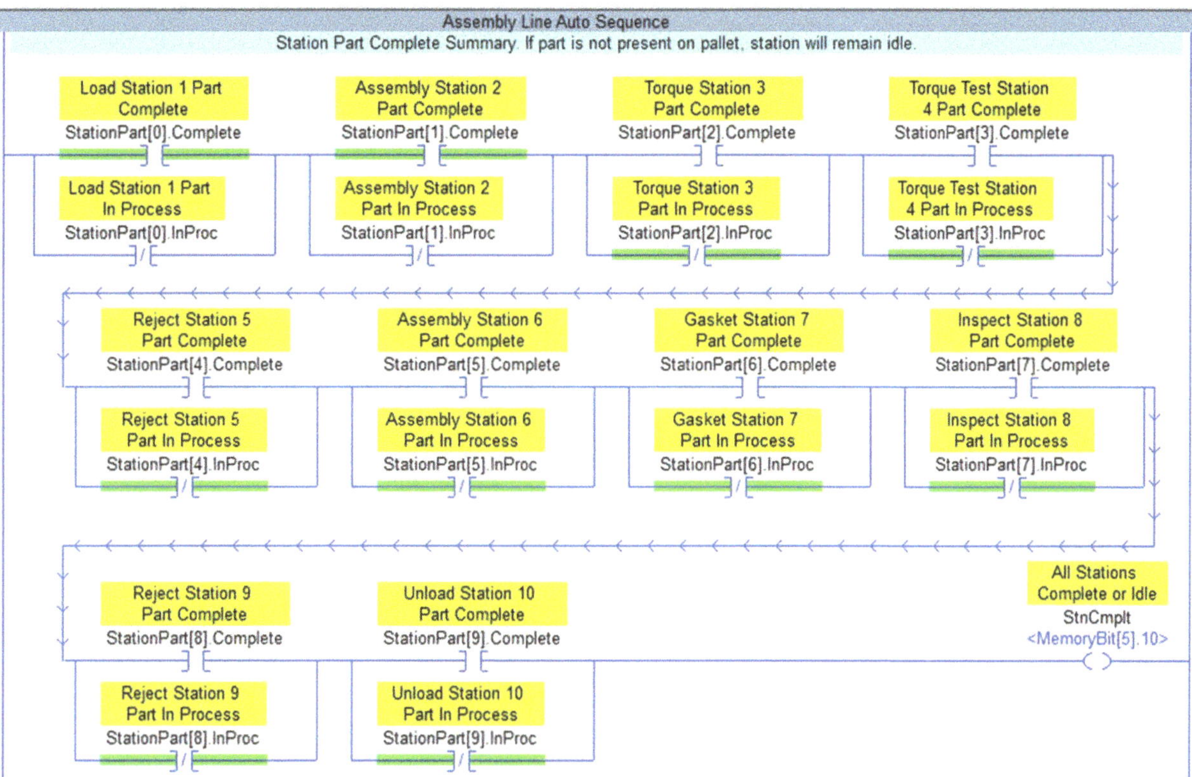

As each part enters a station, the *InProc* bit is set in the corresponding file, indicating whether a part is present on the pallet. A station that is not In Process is considered idle. When the operation at the station is finished, the *Complete* bit is set indicating that it is okay to index the line.

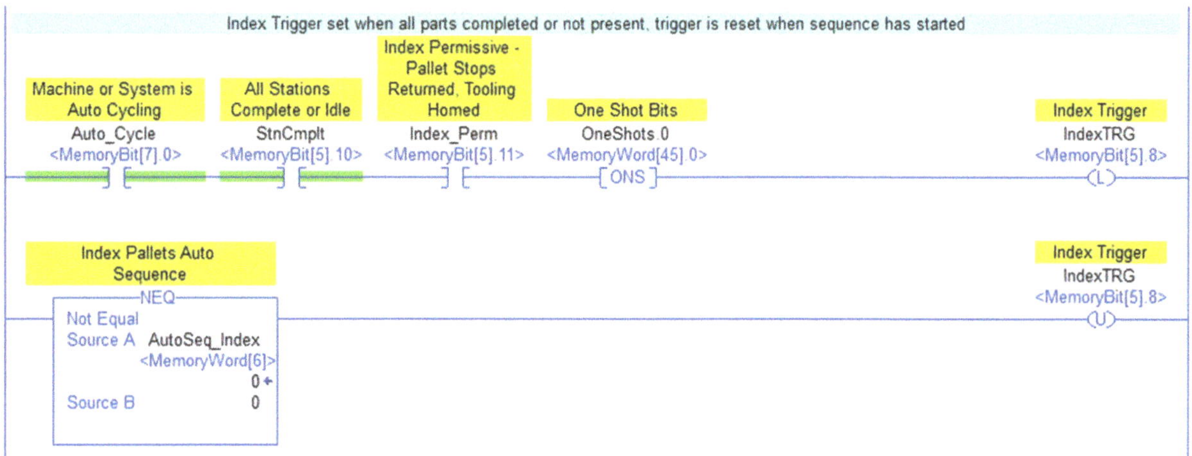

The Index Trigger is used to start the pallet movement sequence when all products' status are correct and when mechanical tooling is in the correct position (*Index_Perm*).

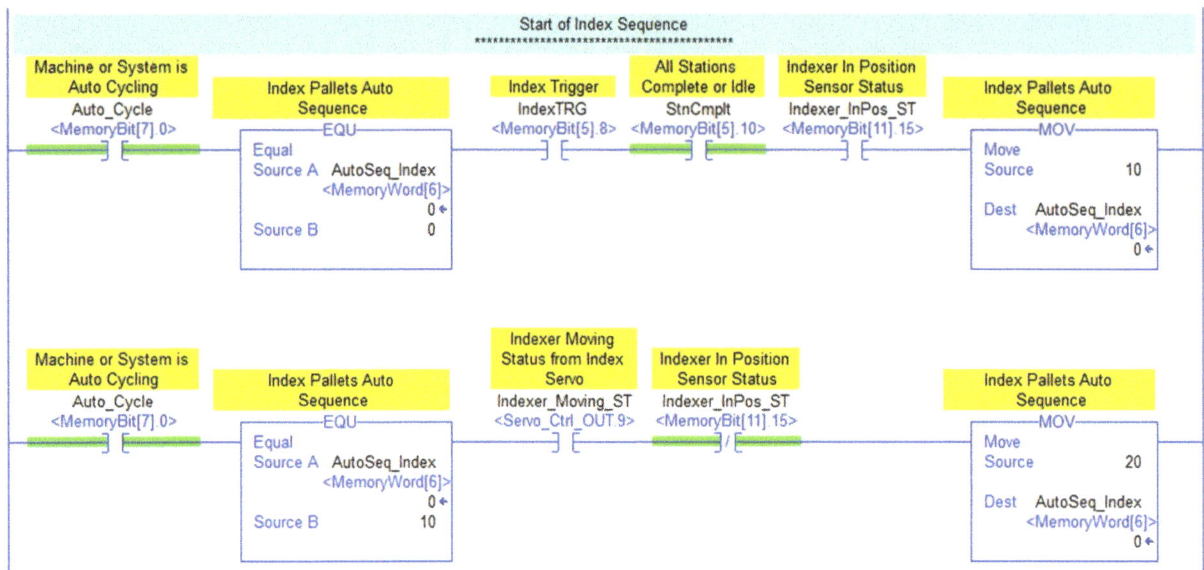

The Index Sequence itself is similar to other Word Sequences discussed in this manual. Step 10 is used to tell a servo type pallet indexer to begin movement. When motion has started, the sequence moves to Step 20, removing the signal to the servo, and when movement is completed, the sequence moves to Step 30. At this time, an *IndexComplete* signal is also set. This bit is used to move the part records.

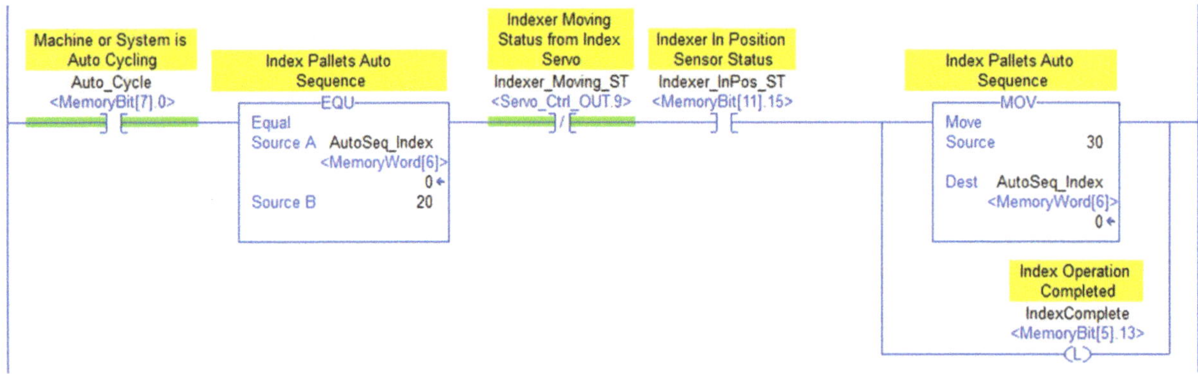

The end of movement also resets all of the Part Complete bits in the part records, as shown with the unlatches on the next page. When all of the part records have been moved in the part tracking routine, the *IndexComplete* bit is reset. This bit serves as a handshake between the servo indexer's movement and the part tracking.

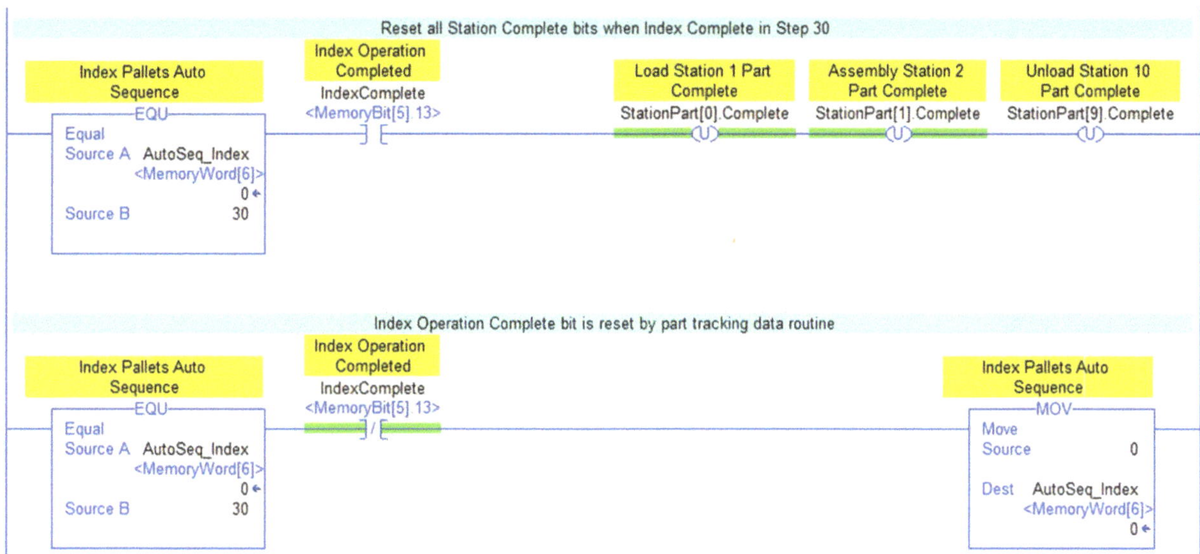

Some PLC brands also only allow data files to be moved one file at a time, while others allow "Block Moves". The MOV instruction will move one file, while the COP instruction will move several.

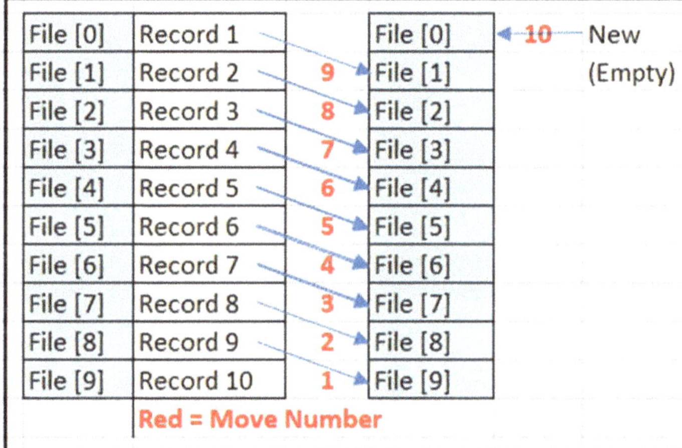

Cascading File Moves: If files must be moved one at a time, it is necessary to move the last part records first so that the newest data is not overwritten. Figure 43 shows a 10-element array, where the 9th record is moved into the 10th spot, the 8th into the 9th spot, and so on. In the final or 10th sequential move, a new empty record is moved into location 1 and is ready to be updated.

The logic for this is sometimes called "cascading" logic, because records cascade from one location to the next.

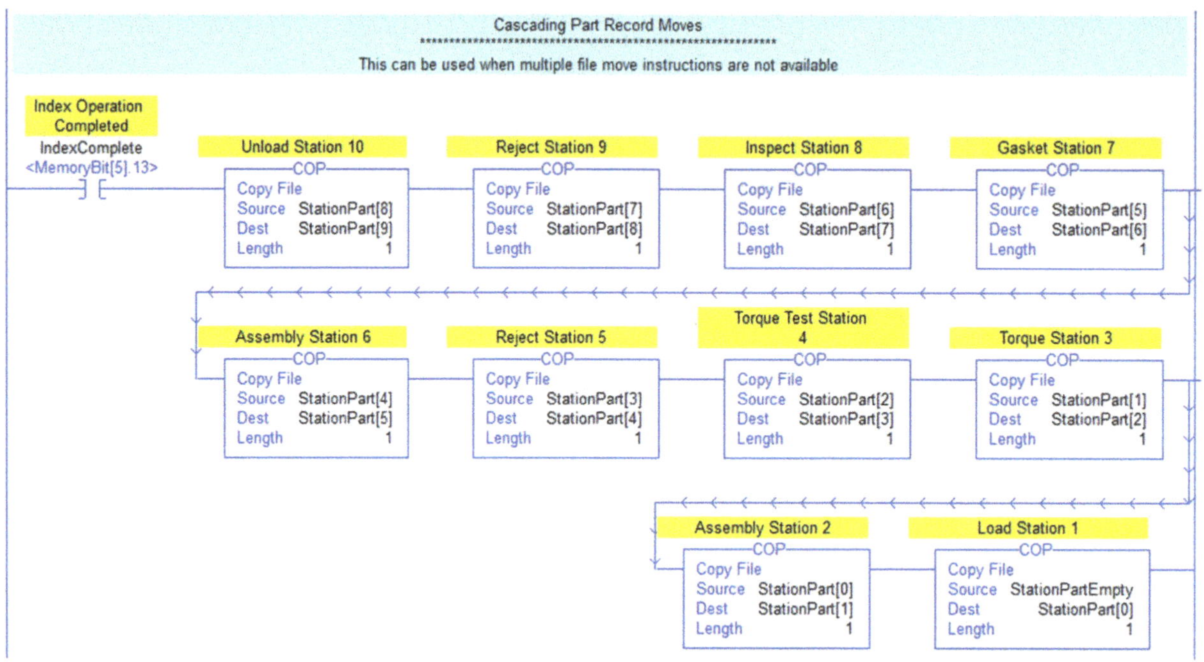

Figure 46 shows individual sequential Copy instructions being used to move data. In the ControlLogix PLC platform, these output type COP instructions can be placed in series. This logic is triggered by the *IndexComplete* signal discussed previously.

Block File Moves: Multiple files can also be moved with the COP instruction. This requires that the data is in array form and also requires an additional "dummy" array the same size as the *StationPart* array.

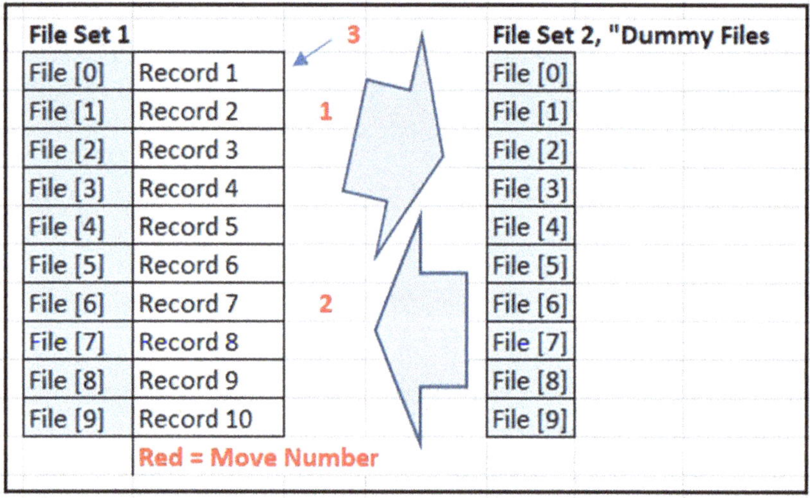

The first operation moves records 1-9 into positions 2-10 in the dummy array. This is because files cannot overwrite themselves on block type moves.

The second operation moves the records in positions 2-10 of the dummy array back into the original array, also into positions 2-10.

The third operation moves the new empty file into position 1 of the record; it is ready to be updated.

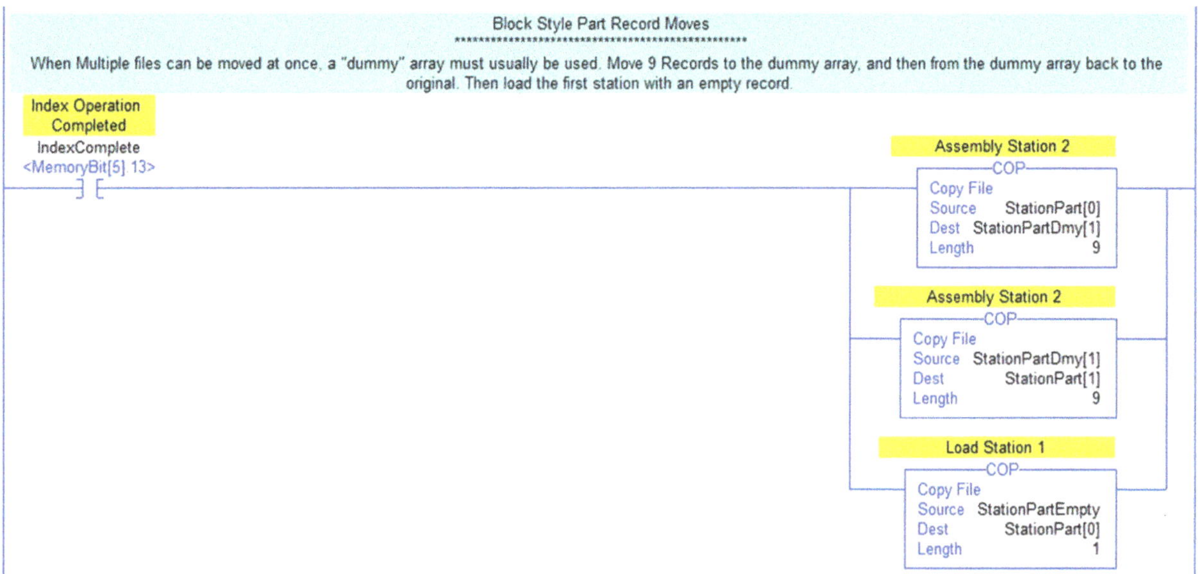

The figure above shows the COP instruction being used to move multiple files at once. A Synchronous Copy (CPS) instruction also exists for the Allen-Bradley ControlLogix PLC that ensures that data can't change while the data is being moved. This Block Move method is preferred if a large number of files or array elements need to be handled, though it does require more memory for the additional "dummy" array.

Following are some additional simple useful methods used in PLC Programming.

Hold-In or Latching Circuit

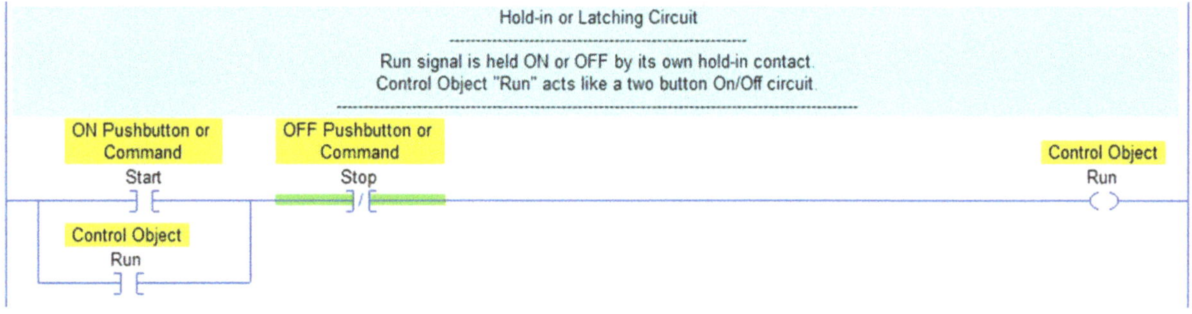

This logic acts as a set and reset pair and is commonly used with various modifications. If the coil address Run's contact is parallel with the logic, it seals in the circuit even if the Start button is not on.

Toggling Pushbutton

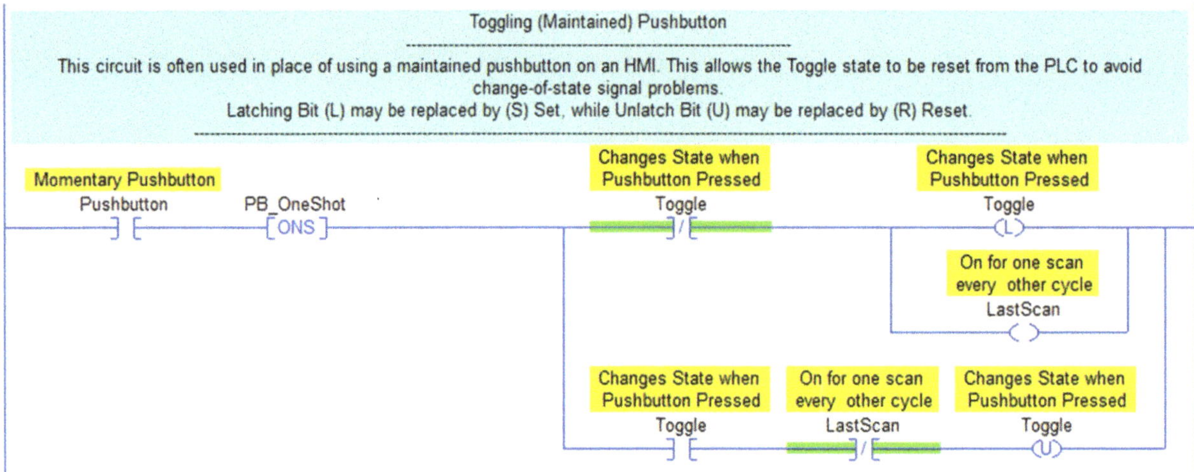

This circuit is used to imitate a maintained or "cammed" pushbutton. When the momentary pushbutton is pressed, it sets the Toggle bit. If it is pressed again, the Toggle bit resets. This can be done in other PLC languages with fewer instructions, as it is essentially an Exclusive OR (XOR) instruction.

"Free Running" Timer

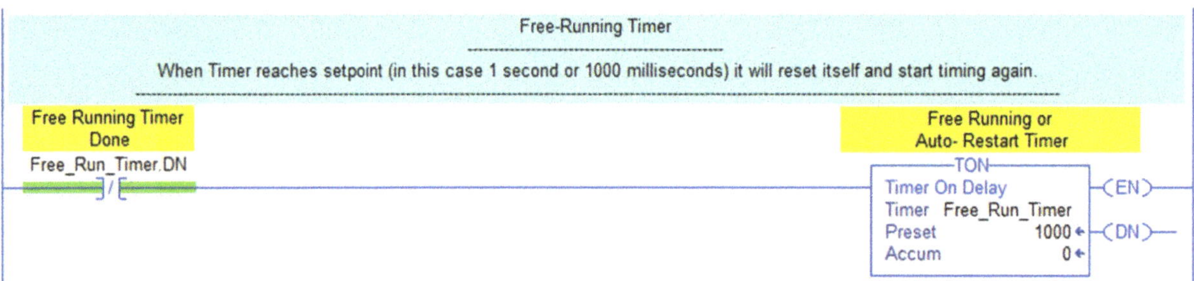

This timer counts up to the setpoint and then resets itself. The accumulator value can be used in many applications, such as creating ramps, and various math functions can be applied. The bits of the accumulator can be used to flash lights at various frequencies.

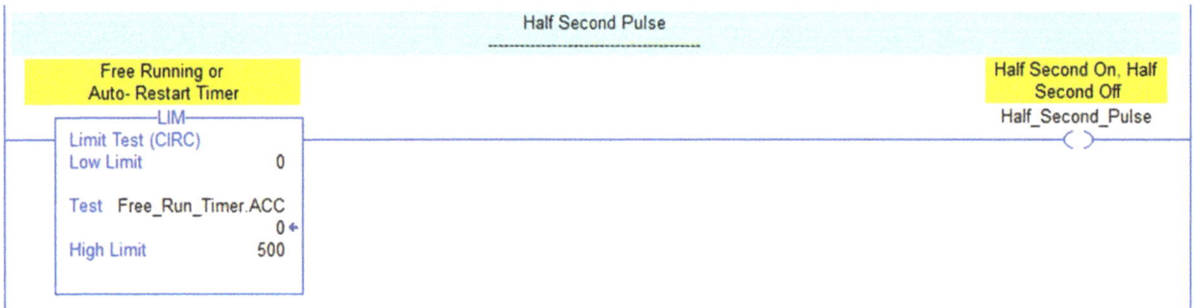

Blink Timers

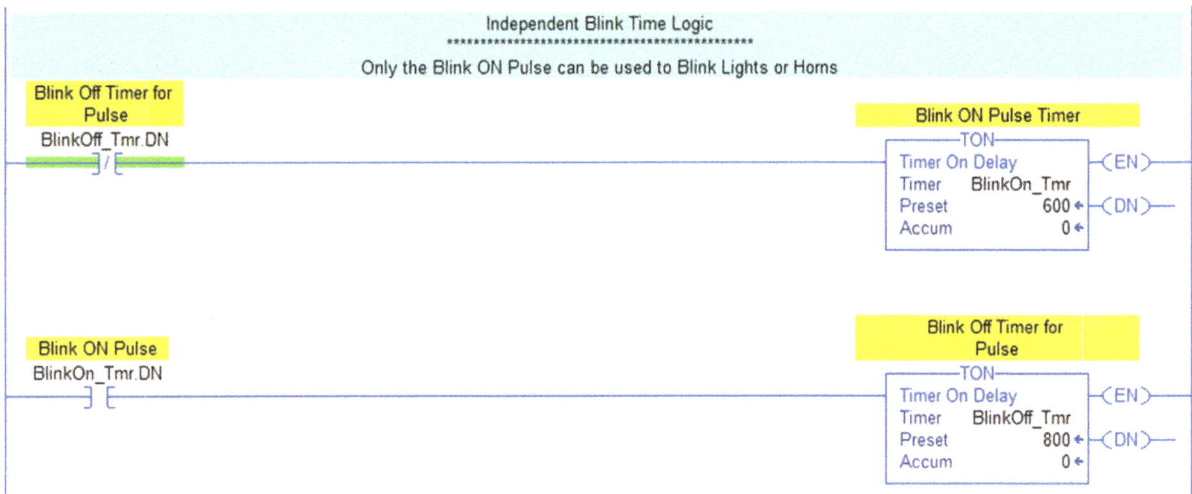

A Blinking Timer pair can be used when specific on and off pulse times are desired. Only the first timer's Done bit can be used; the Blink Off timer's Done bit is a one-shot.

Accumulators and Decumulators

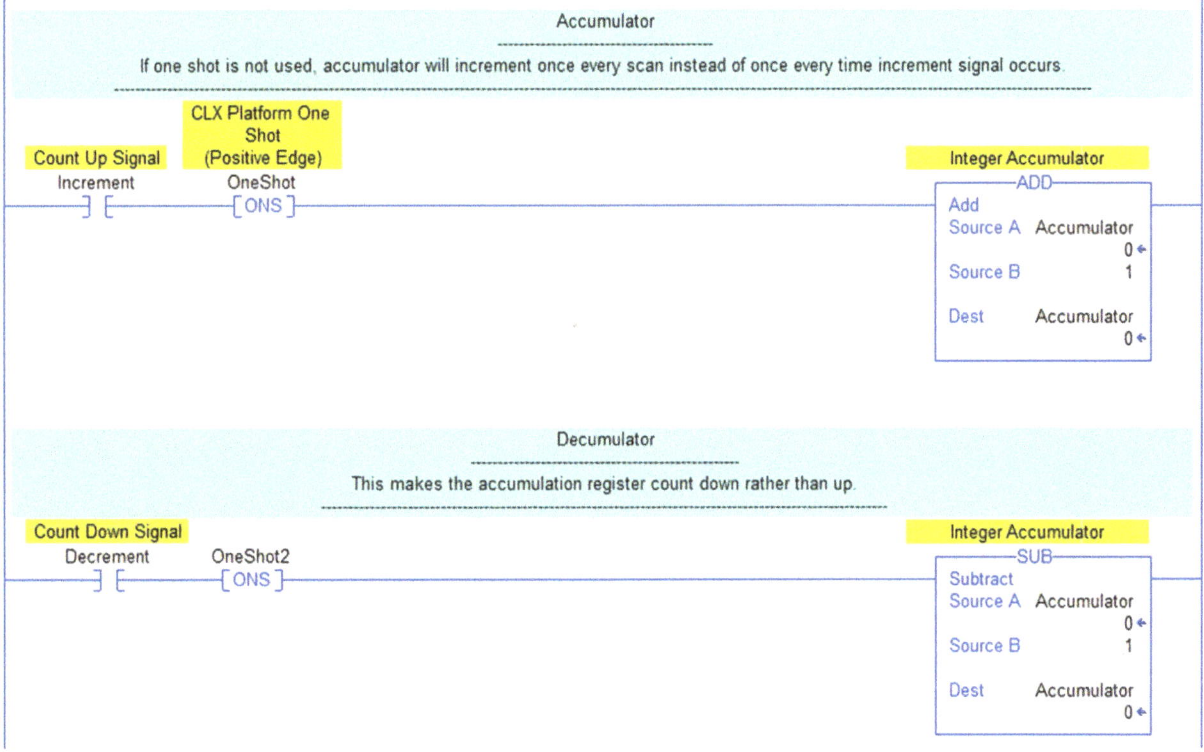

A number can be added to a register and then placed back into itself, as shown above. These rungs act like an Up-Down Counter. Some PLC platforms have limitations, such as a low maximum setpoint value, the done bit changing state at zero rather than at the setpoint, or they use BCD instead of decimal numbers. This logic allows the programmer to build their own counter.

To create a done bit, use a Greater Than or Equal Instruction on the value; to reset the counter, move a zero into the value.

Simulation

For software simulation, routines are created that interact with the code that the programmer writes. If an actuator is simulated, when the output controlling the actuator is activated, a timer starts with a setpoint approximating the time it takes the actuator to reach a sensor. The simulated input is activated by the Done bit of the timer. If the actuator has analog positional feedback, a ramping accumulator can be used to add to a value over a time period (combining the free-running timer with the accumulator/decumulator in the previous section). This kind of logic can be used to step an auto sequence through its stages.

Most newer HMI software also has the ability to "**animate**" objects and show them moving to different positions. Rather than using physical models, a picture of the actuators can be used to show the status of the machinery.

Simulation is often used by programmers to test the links between HMI objects, such as pushbuttons and indicators and the PLC program. This is easier than testing the auto sequence logic as described here. Bits and numbers can simply be modified in the program to change the HMI display.

Writing code to simulate the entire program is much more complicated than simply testing links between objects. It requires mapping physical I/O to simulated I/O, since the physical inputs can't be modified. Input feedback needs timers and ramps for every point used in the program. This code should be kept in separate routines to isolate it from the operational part of the program; it can always be deleted or deactivated later. Outputs can even be used directly as long as they aren't connected to operating equipment!

In the following program examples, simulated input and output registers were created as an array of DINTs called Sim_DigInput[x] and Sim_DigOutput[x], where [x] is the element number of the array.

In many ControlLogix PLC templates, descriptive tagnames are aliased to I/O addresses as a standard. In this simulation, local (program) descriptive tags are aliased to simulation tags as shown below:

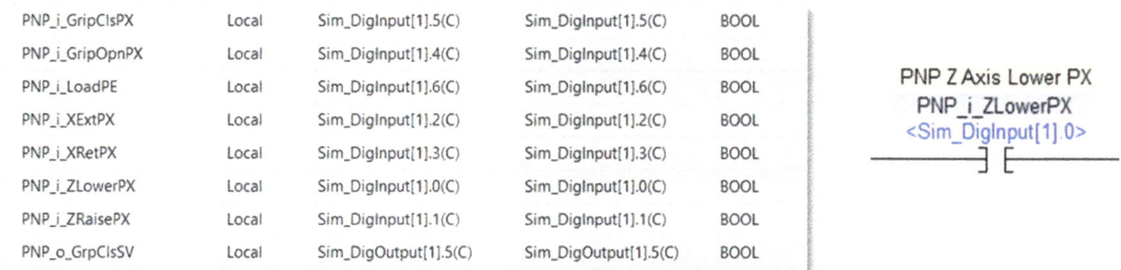

This allows the alias to be changed to a real (physical) I/O address (such as Local:2:I.Data[2].12) after code has been tested.

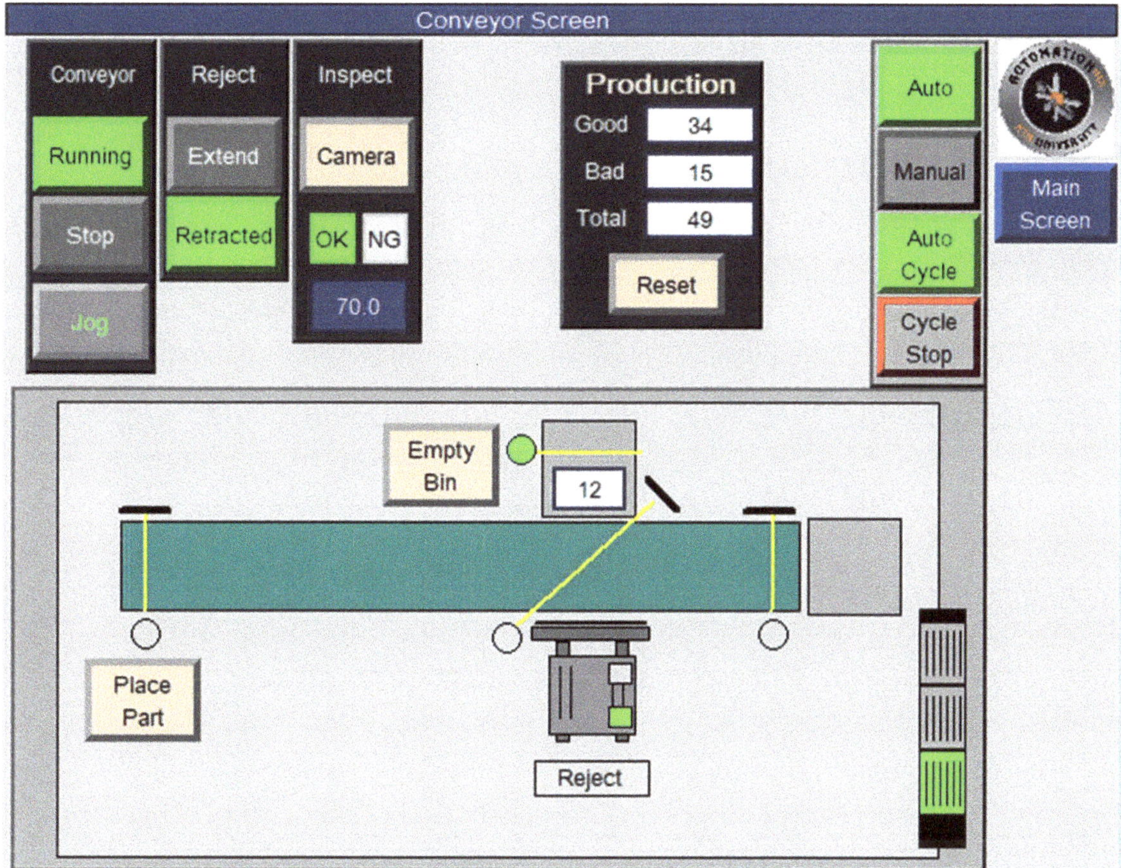

The HMI screen above shows an image of a conveyor with controls for a reject mechanism and the conveyor itself. There are buttons for mode control and an imitation stack light. This is similar to the trainer used for the programming exercises for Automation NTH's 103 PLC course.

There are two small square sensor indicators on the reject mechanism, these are animated by the tags Reject_i_ExtPX and Reject_i_RetPX, which in turn are aliased to two of the Sim_DigInput tags. The outputs that control the reject mechanism are aliased to Sim_DigOutputs.

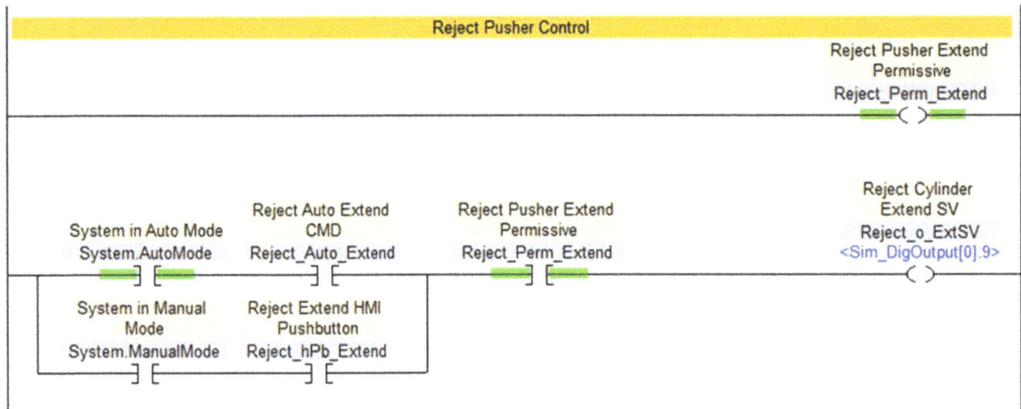

The outputs for the Reject cylinder are controlled in the usual way, with a pushbutton in Manual Mode and with a CMD bit in Auto. Instead of a physical output, the Sim_DigOutput bit will be activated.

⊿ ♭ MainProgram
  ◇ Parameters and Local Tags
  ▣ MainRoutine
  ▤ r04_PulseGenerator
  ▤ r10_System
  ▤ r11_Inputs
  ▤ r12_AutoSeq
  ▤ r13_Outputs
  ▤ r14_Faults
  ▤ r15_PartTracking
  ▤ r20_Production
  ▤ r30_HMI
  ▤ r90_Sim
  ▤ r91_Examples

The figure to the left shows the call structure for the control of this simulation. r12_AutoSeq contains the code for activating the Reject_Auto_Extend command tag, which will be activated if a rejected part is detected.

R90_Sim contains all logic for making the sensors and simulated movement of objects happen. There are several AOIs that have been developed to make this easier.

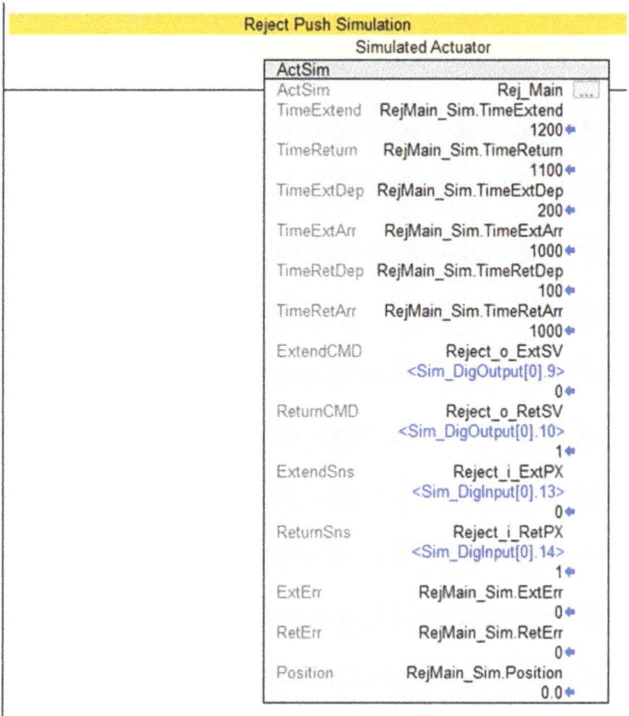

The **ActSim** AOI allows linear actuators to be simulated. A UDT is created for the actuator of type u_SimAct, it has the same basic elements as the AOI itself. The AOI contains timers for extending and retracting the cylinder. When the ExtendCMD or ReturnCMD tags are activated, the corresponding timer will run as long as its Err bit is not on.

The numbers in TimeExtend and TimeReturn are the time it will take for full extension and retraction of the cylinder in milliseconds. The ExtDep and RetDep values are the time it takes for the actuator to leave the corresponding sensor, and the ExtArr and RetArr values are how long before the cylinder reaches full travel that the sensor will come on. The Position value is a REAL number that ranges from 1-100, it represents the movement of the actuator as a percentage.

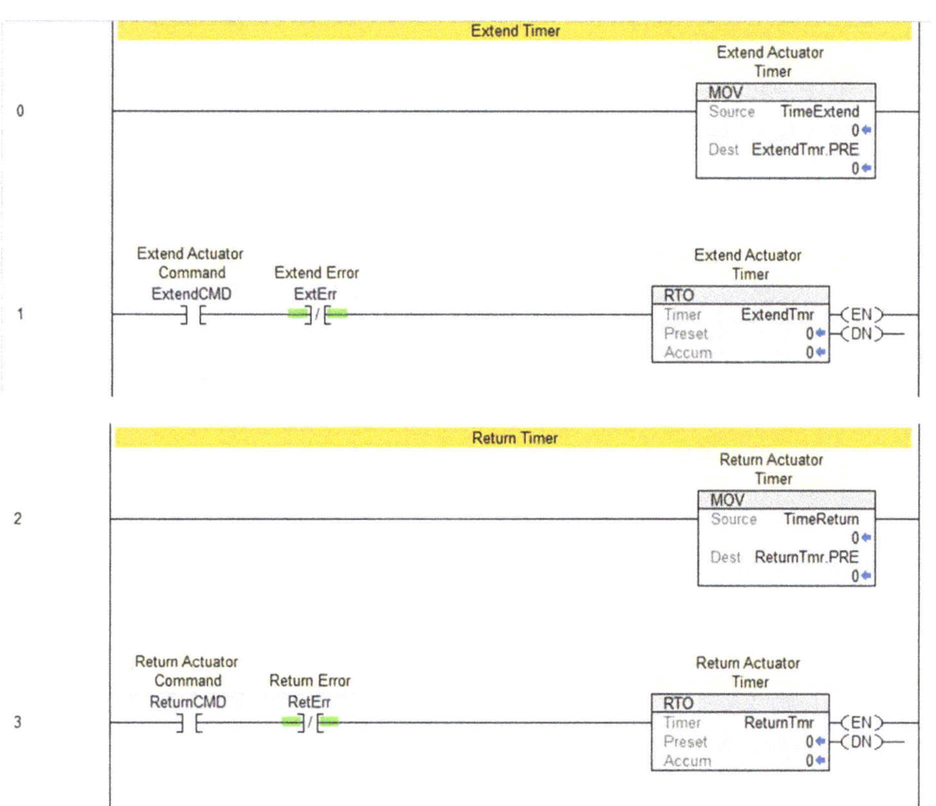

Inside AOI:

The Extend and Return timers are retentive On-Delays that track how long the actuator has been on. The presets TimeExtend and Time Return have to be passed into the timers. ExtErr and RetErr are bits that allow a fault such as an obstruction or bad solenoid valve to be simulated.

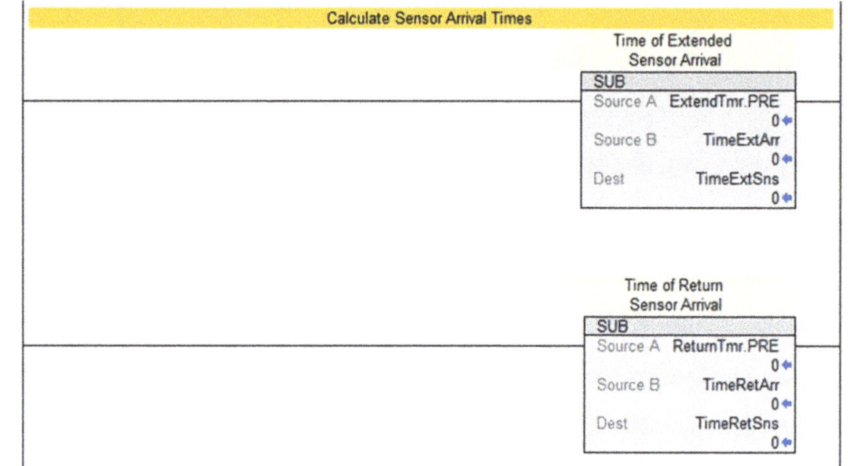

Arrival times for Extend and Return are calculated. The departure times are from zero.

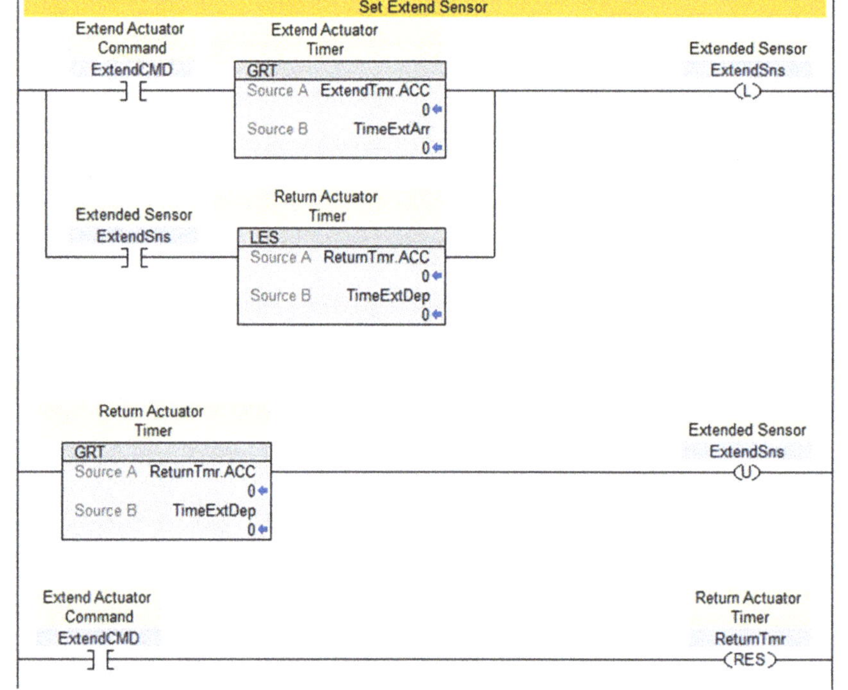

Status of the Extended Sensor is set based on the timer's accumulated value compared to the TimeExtArr setpoint and the TimeExt Dep value. The Return timer is reset when the extend CMD is issued.

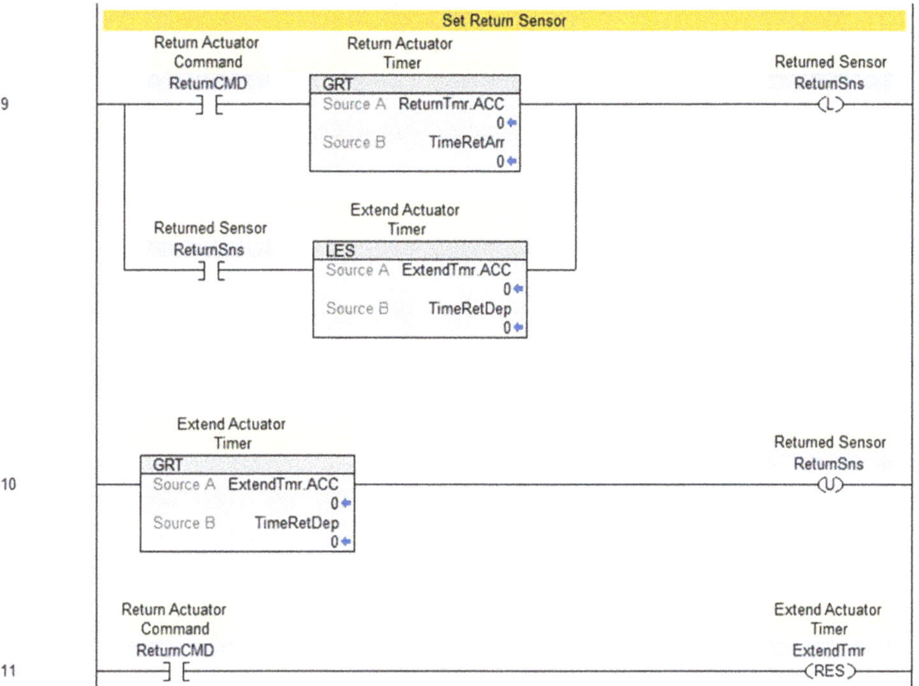

The Return Sensor status is determined in the same way as the Extend Sensor.

The Extend Timer is reset when the Return CMD is issued.

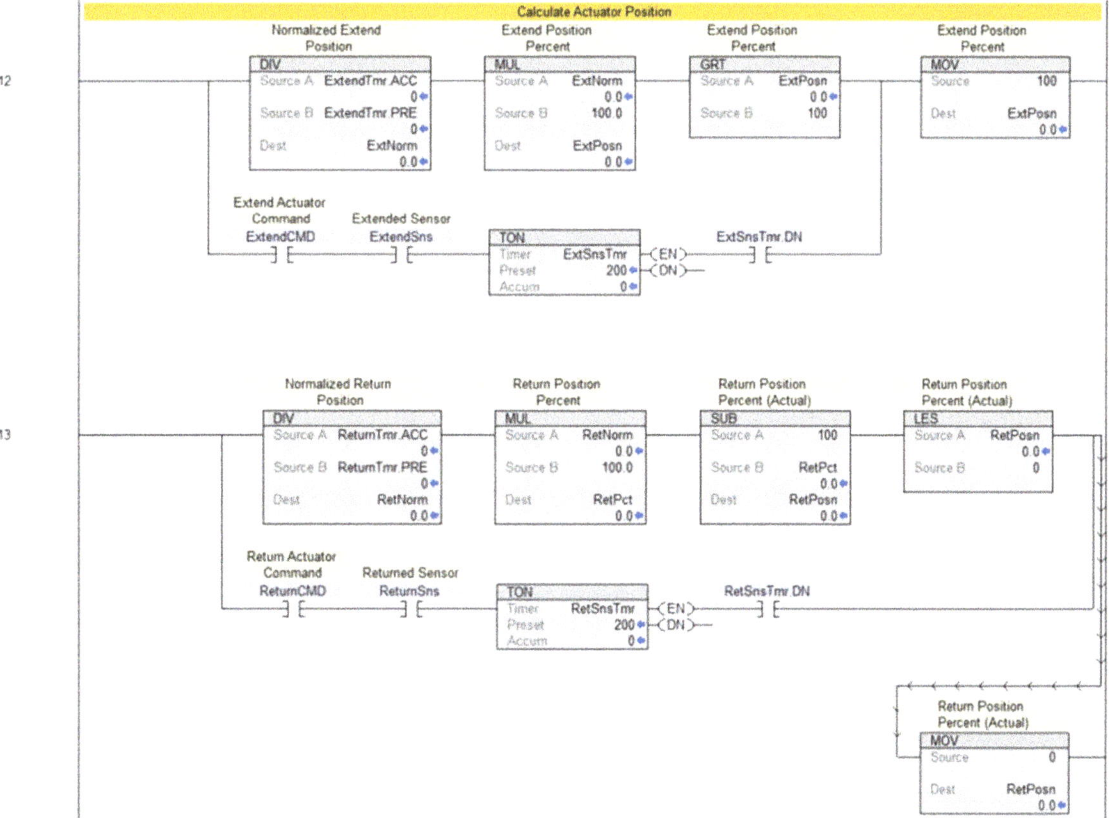

Actuator position is calculated as shown. The two timers ensure that the values are correct when the sensors are active.

The last two rungs in the AOI move the normalized Extend and Return values into the Position tag based on which direction the actuator is moving.

The visibility of the part on the conveyor is also determined in the r90_Sim routine. Pressing the Place Part button on the HMI sets a bit that controls the property for visibility in the Animation tab for the object.

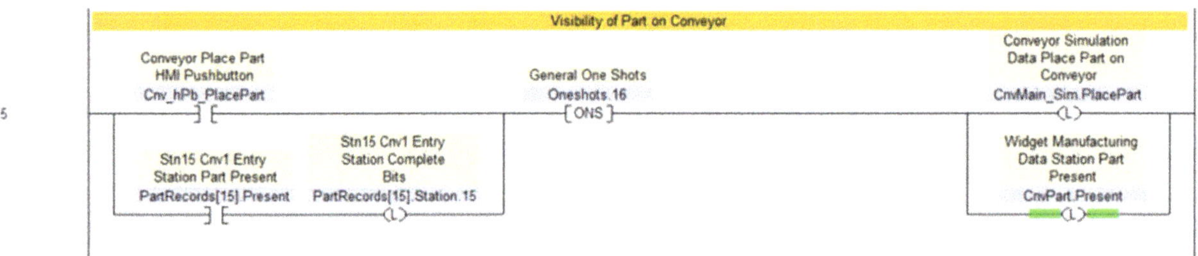

When running the simulation in Auto Mode the part record itself (Part Tracking – PartRecords[15].Station.15) will make the part appear.

The part can't be removed until it leaves the Part Present photoeye at the entry station.

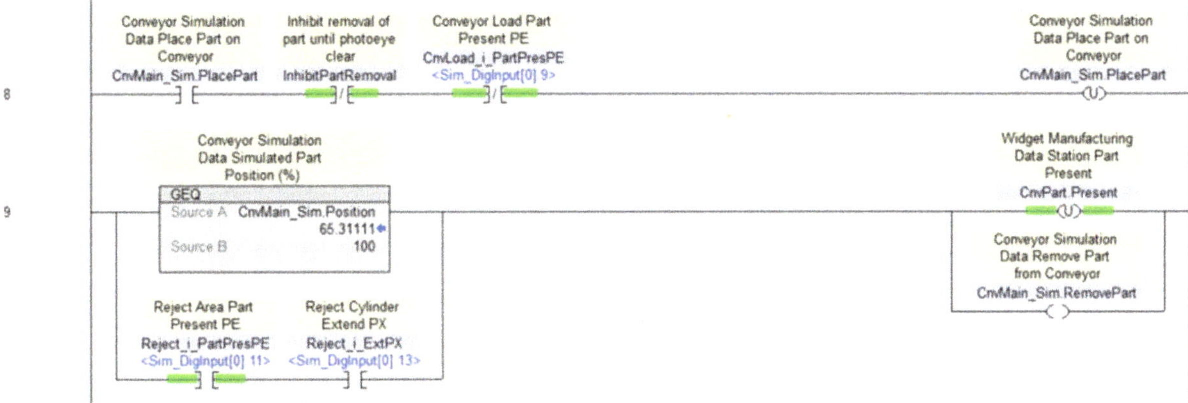

Place Part status is unlatched when the part leaves the load area photoeye, while the visibility of the part is unlatched based in the position of the part on the conveyor, or on the reject mechanism.

The Conveyor also has an AOI that controls the visibility and position of the part.

Inside AOI:

The CnvSim AOI controls the status of photoeyes at each end of the conveyor as well as the position of the part along the conveyor.

The conveyor can also be treated as a Variable Frequency Drive, setting the SpeedCMD tag at a value between 1 and 100% scales the positional movement speed. The RunCMD and StopCMD tags can be set as the same tag for two wire control or to different tags for three wire control. The StopCMD acts as an Enable and must be on for the conveyor to run.

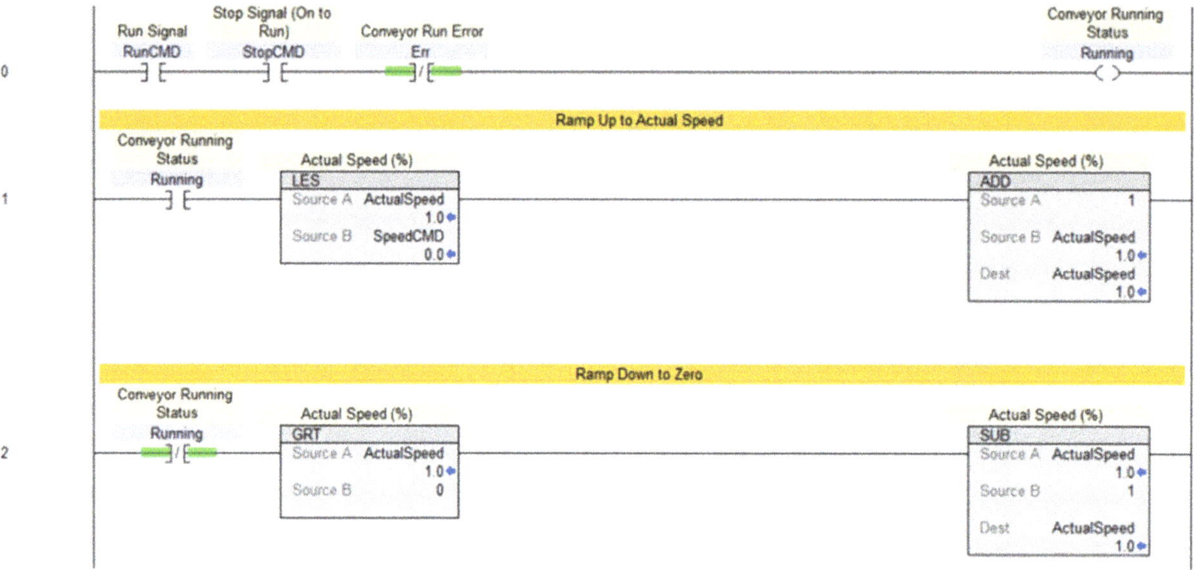

The running status of the conveyor ramps the ActualSpeed output tag value up and down. This could be scaled with additional accel or decel tags if the AOI was modified.

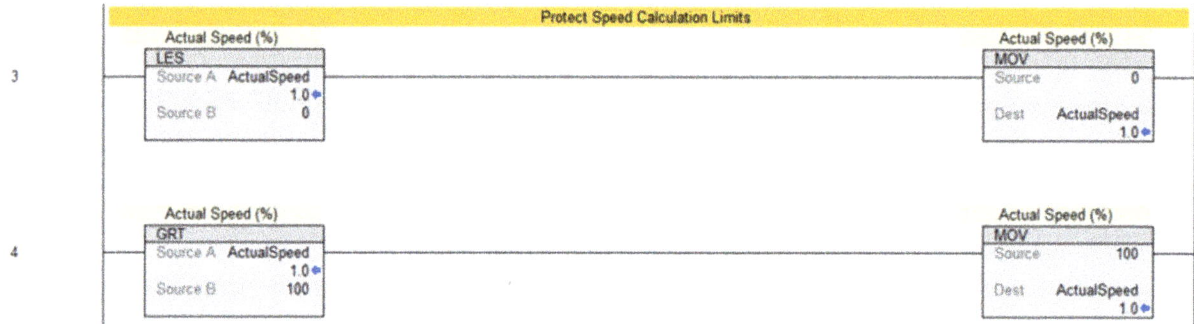

Upper and lower speed settings are confined to a range of 0-100%.

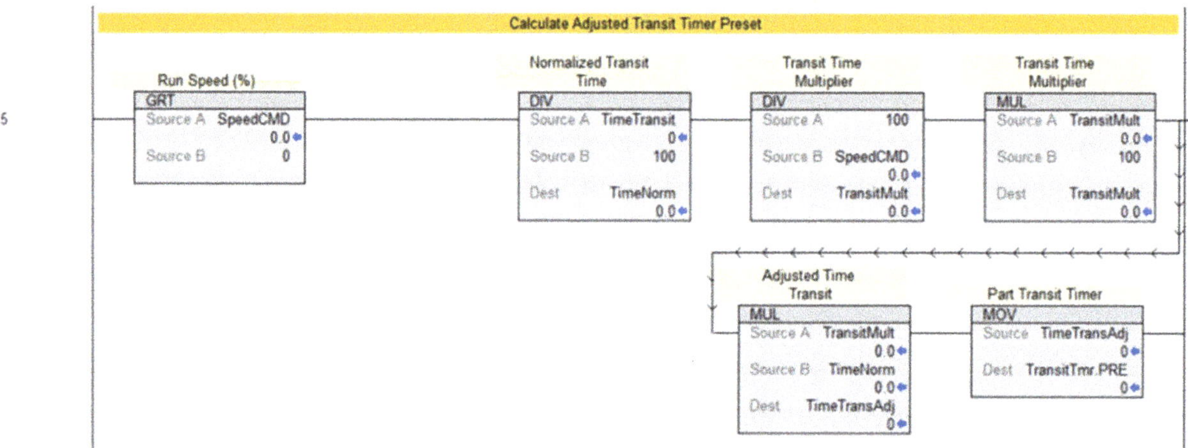

The transit time of the part is scaled by the Speed CMD tag.

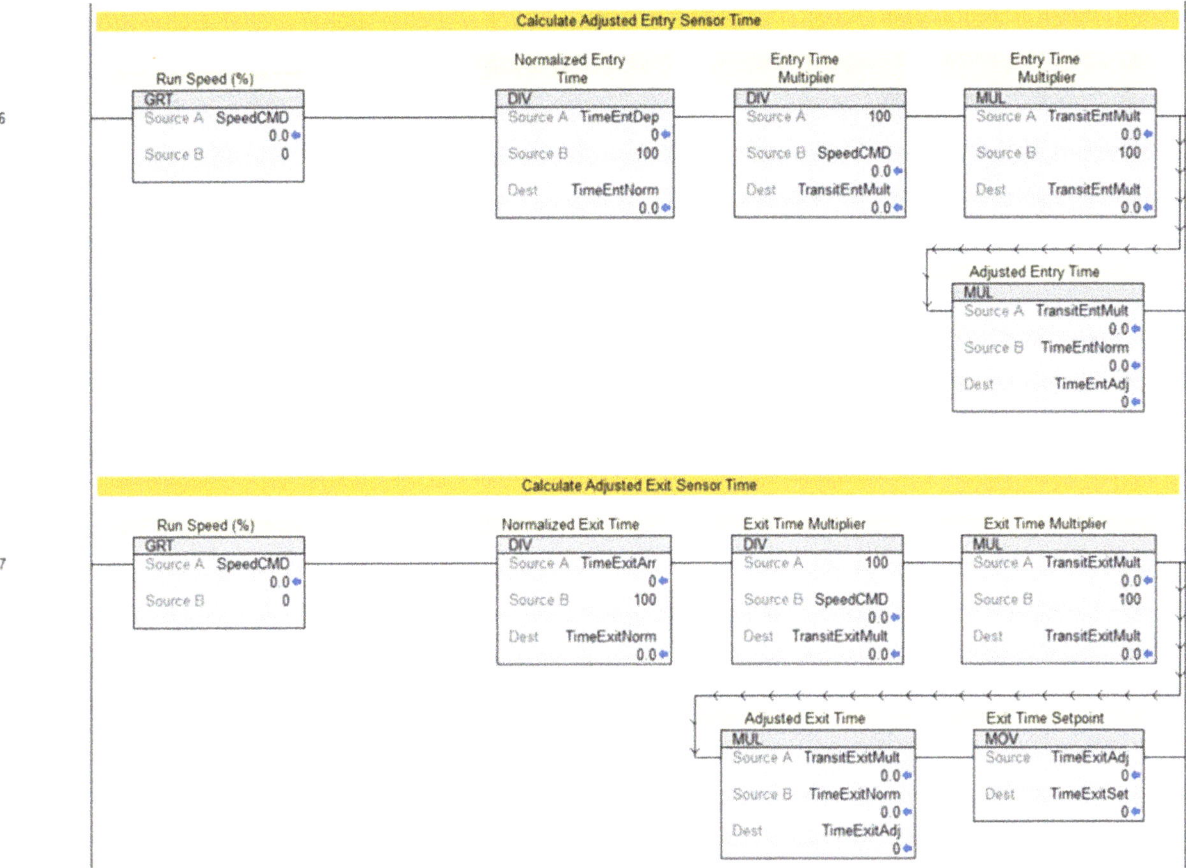

The timing for both the entry and exit photoeye activation is also scaled by the SpeedCMD tag.

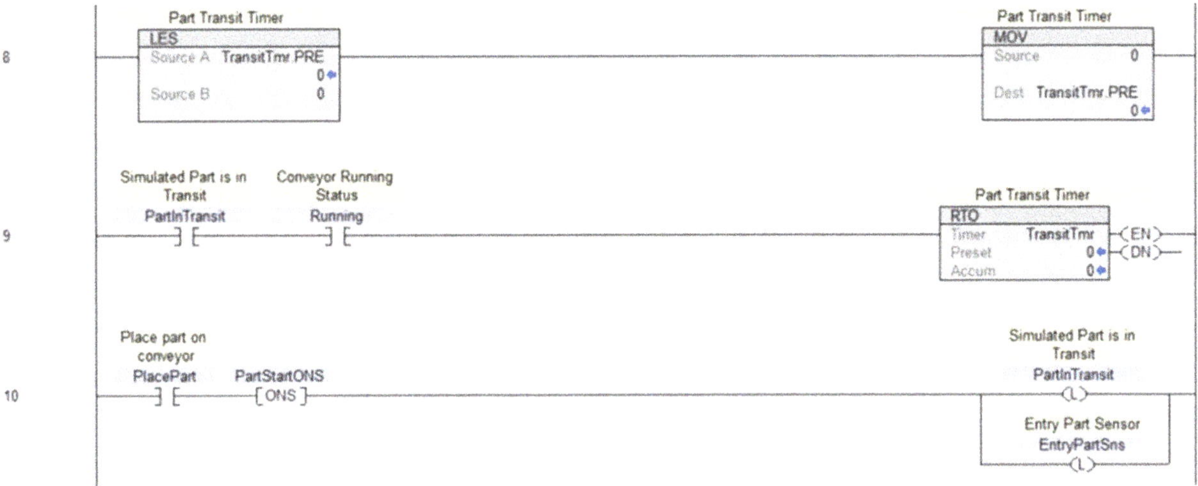

The lower limit of the transit time is protected, and an RTO is used to track the position/time of the part on the conveyor. The PlacePart tag establishes the presence of the part on the conveyor and the status of the entry photoeye.

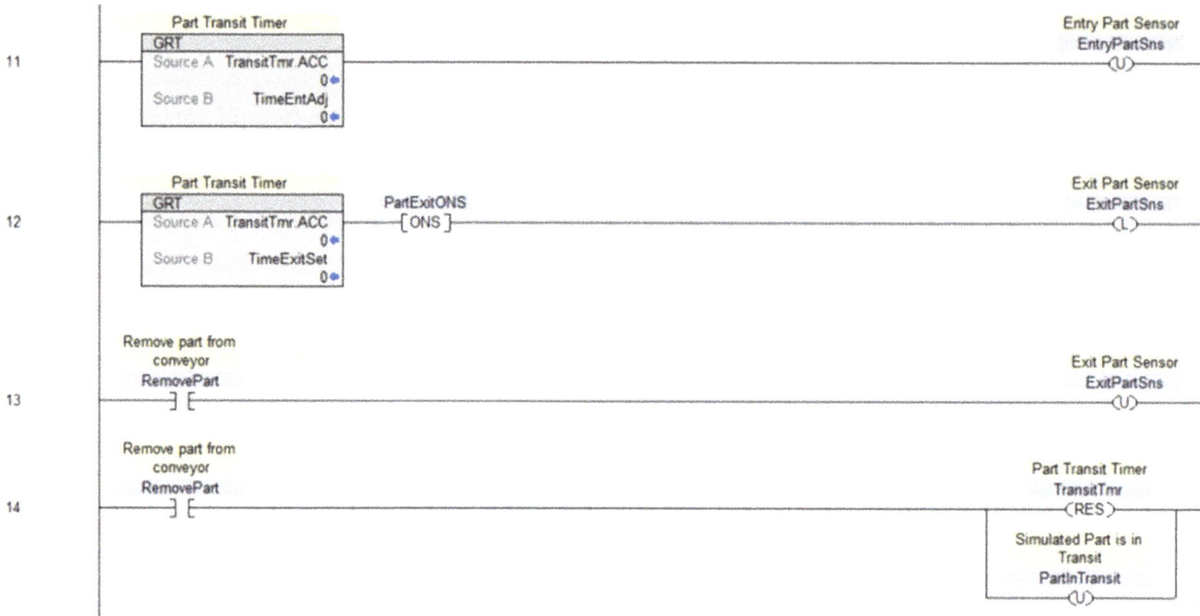

The entry sensor status is reset based on the Transit Timer value. The exit sensor status is controlled in the same way. The RemovePart signal resets the timer and allows the AOI to be used for the next part.

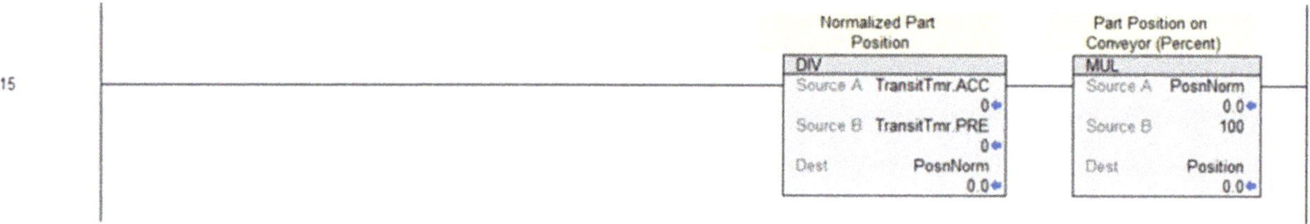

The final position output is scaled and turned into a percentage here. This Position value is used to show the part in an animation on the HMI.

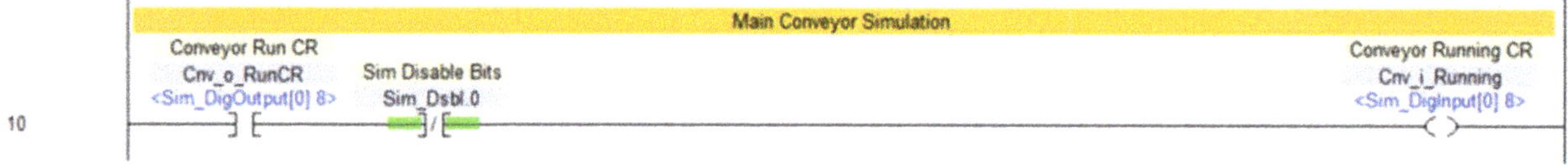

In the r90_Sim routine, the Conveyor Running signal is mapped to the Run command output. To simulate a fault in the drive, the Sim Disable bit can be toggled. This technique allows fault detection and messaging to be tested.

Summary

This book covers many of the products and techniques used in programming Allen-Bradley PLCs. In the NTH University internship/apprenticeship program, students write programs in both RSLogix500 and Studio 5000 to control a simple conveyor with reject mechanism as shown below; they are held to a high standard in ensuring that the system operates correctly and that the code is well-written. Writing the PLC and HMI programs and testing them takes several weeks.

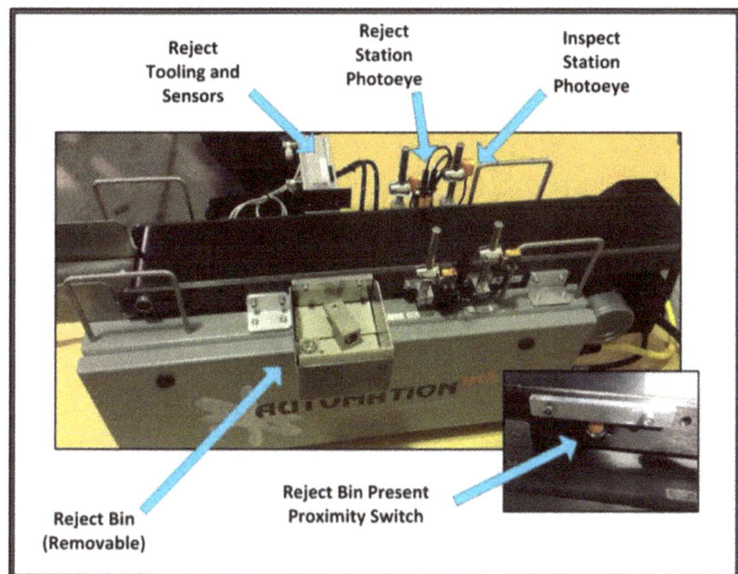

They also spend time designing control panels with CAD and working in the panel shop. After completion they attend a final class taught by the author where they program a simulation building on the program and using the AOIs shown in this book. They are also introduced to a template used by the company to program large complex machines.

Learning PLC programming is only part of a much larger field of study for controls engineers. It is the start of a journey that will take many years and often involve learning other PLC platforms and HMI/SCADA interfaces.

Hopefully this book will provide a foundation for continuing this journey and developing excellence in your career!

About the Author

Frank Lamb is an industrial automation consultant and advanced PLC programming trainer with more than 30 years of experience in controls and machine automation.

From 1996 to 2006, Frank owned and operated Automation Consulting Services, Inc. (ACS), a panel building and machine integration company in Knoxville, TN. From 2006 to 2011, he worked as a senior-level project engineer for Wright Industries in Nashville, TN, where he led the design and implementation of large, complex systems and custom machines for multinational corporations and government agencies. In December 2011, Frank re-established Automation Consulting, LLC with a new vision: to use his experience in the field of industrial automation – from electrical, mechanical and controls engineering to project management, training, and machine documentation – to provide expert consulting and training services to manufacturers.

Frank is the president and owner of Automation Consulting, LLC in Nashville, TN and works as Lead Trainer for Automation NTH in LaVergne, Tennessee. He is the author of several books including *Industrial Automation: Hands On*, published by McGraw-Hill Professional in 2013, *Advanced PLC Hardware and Programming*, published by Automation Consulting, LLC in 2019, and *Maintenance and Troubleshooting in Industrial Automation*, published by Automation Consulting, LLC in 2022. He is a United States Air Force veteran, received his BSEE in Electrical and Computer Engineering from the University of Tennessee, and has a Green Belt in Lean Manufacturing/Six Sigma from Purdue University.

www.ingramcontent.com/pod-product-compliance
Lightning Source LLC
Chambersburg PA
CBHW040143110726
48005CB00018B/2627